Geology of the Sc Coalfield, Part V. around Bridgend

This memoir describes the geology of the district around Bridgend, largely situated to the south of the South Wales Coalfield and forming the western part of the Vale of Glamorgan. It explains the sequences of rocks and superficial deposits, encompassing a geological history of some 395 million years.

The rocks comprise limestones, sandstones and mudstones of Devonian and Carboniferous age, disposed in a major fold, the Cardiff–Cowbridge Anticline. They are overlain by mostly red to green marls and sandstones of Triassic age and thinly interbedded limestones and shales of the Lower Jurassic. The latter form the spectacular cliffline of the Glamorgan Heritage Coast. The superficial deposits of the last glaciation, which are preserved in the north and east of the district, are also described, together with those developed in the river valleys and along the coast during the postglacial period. A chapter on the tectonic evolution and mineralisation of the district is included, together with an economic geology chapter dealing with the main resources, including water supply. A reference list, summary logs of key boreholes and an excursion guide to the coast near Ogmore-by-Sea are also provided.

The Blue Lias of the Glamorgan Heritage Coast looking south from Dunraven towards Whitmore Stairs. The cliffs of alternating limestones and shales comprise Units A and B of the Porthkerry Formation (Photograph by M Smith)

BRITISH GEOLOGICAL SURVEY

D WILSON, J R DAVIES,
C J N FLETCHER and
M SMITH

Geology of the South Wales Coalfield, Part VI, the country around Bridgend

Memoir for 1:50 000 geological sheet 261 and 262
(England and Wales)

SECOND EDITION

LONDON: HMSO 1990

First published 1904
Second edition 1990

ISBN 0 11 884425 3

Bibliographical reference

WILSON, D, DAVIES, J R, FLETCHER, C J N and SMITH, M. 1990. Geology of the South Wales Coalfield, Part VI, the country around Bridgend. *Memoir of the British Geological Survey*, Sheet 261 and 262 (England and Wales).

Authors

D Wilson, BSc, PhD
J R Davies, BSc, PhD
C J N Fletcher, MSc, PhD, C Eng. MIMM
M Smith, BSc, PhD
British Geological Survey, Aberystwyth

Other publications of the Survey dealing with this district and adjoining districts are given in Appendix 3.

Printed in the UK for HMSO

Dd 240425 C10 5/90

CONTENTS

TABLES

FIGURES

PLATES

PREFACE TO THE SECOND EDITION

This memoir describes the geology of the district covered by 1:50 000 sheet 262 (Bridgend), including a small land area on adjacent sheet 261 (Sker Point). It was originally surveyed on the one-inch scale, on Old Series sheets 20, 36 and 37, by H T De la Beche, W E Logan, D H Williams, W T Aveline and T E James, and published in about 1845; a revised edition, incorporating work by H W Bristow and H. B. Woodward, was published in 1873. The first survey on the six-inch scale was by A Strahan, T C Cantrill and R H Tiddeman, the results being published as a one-inch sheet in 1901, with an accompanying memoir in 1904. The Silesian rocks of the district were partly resurveyed by A W Woodland between 1948 and 1953, during the mapping of the adjacent Pontypridd sheet, and published as 1:10 560 County Series sheets in 1957 and 1960. These maps were partly revised by I H S Hall and H C Squirrell and published as reconstituted 1:10 560 National Grid sheets between 1966 and 1971.

A narrow strip of land along the eastern margin of the sheet was resurveyed by Mr K Taylor and Dr R A Waters between 1976 and 1977 as part of the survey of sheet 263 (Cardiff), but the latest survey, on the 1:10 000 scale, was mostly undertaken as a contract for the Department of the Environment on behalf of the Welsh Office, by Dr C J N Fletcher (Project Leader), Dr J R Davies, Mr M Smith and Dr D Wilson between 1981 and 1983. The latter was responsible for the compilation of this short memoir and the work was carried out under the overall supervision of Dr R A B Bazley as Regional Geologist.

Identification of fossils collected during the recent survey was by Miss D M Gregory (Quaternary), Dr H C Ivimey-Cook (Jurassic), Mr M Mitchell (Dinantian), Dr N J Riley (Silesian) and Dr A R E Strank (Dinantian). Photography was mainly by Mr C J Jeffrey. Mrs M A Lewis provided information on the hydrogeology.

Grateful acknowledgement is made to British Coal for providing access to borehole and mining data and, in particular, to Mr C R Thewliss for his assistance with interpretation of the Coal Measures. Acknowledgment is also made to the National Museum of Wales for providing access to borehole collections, and to the various local and county authorities for information and assistance. The willing co-operation of landowners and quarry owners, at the time of survey, is also greatly appreciated.

F G Larminie, OBE
Director

British Geological Survey
Keyworth
Nottingham
NG12 5GG

12 February 1990

PREFACE TO THE FIRST EDITION

The country around Bridgend, which is described in this the Sixth Part of the Memoir on the Geology of the South Wales Coal Field, is illustrated in Sheet 261–2 of the New Series One-inch Map. The area is occupied for the most part by Lower Lias, and by the Rhaetic and Keuper subdivisions of the Trias, but it includes a small part of the South Crop of the Coal Field and numerous inliers of Carboniferous Limestone which project through the Secondary strata.

The original survey was made by Sir H. T. De la Beche, Sir W. E. Logan, D. H. Williams, W. T. Aveline, and T. E James on the Old Series One-inch Maps 20, 36, and 37, which were published about the year 1845. The separation of the Rhaetic strata and a revision of the area south of the Coalfield were begun by H. W. Bristow in 1864, and completed by Mr. H. B. Woodward in 1871. The revised edition was published in 1873.

The re-survey was made on the six-inch scale under the superintendence of Mr. Strahan and was published in 1901. The western half of the map was surveyed by Mr. Tiddeman, the eastern half by Mr. Cantrill, with the exception of the north-eastern margin which was surveyed by Mr. Strahan. The present volume includes the description of the eastern half by Mr. Cantrill and of the western half by Mr. Strahan, based on the notes left by Mr. Tiddeman on his retirement from the staff. Mr. Woodward supplies an account of the Liassic and Rhaetic tracts which were examined by him for the purposes of the General Memoir on the Lias of England and Wales. The principal changes made in the map as the result of the re-survey consist in the extension of the Lias of Cowbridge over a district formerly thought to be occupied by Old Red Sandstone, the discovery of further inliers of Carboniferous Limestone in the Lias near Wick, and the distinction of the litoral from the normal type of Lias throughout the whole area.

The Old Red Sandstone includes, in its upper part, some quartzitic and pebbly grits; the whole, so far as exposed, appears to correspond in character to the Brownstone subdivision, but in the absence of fossils it cannot be definitely referred to either Upper or Lower Old Red Sandstone.

The Carboniferous Limestone maintains the same general character as in the Cardiff district. Thick but impersistent limestones appear locally in the Lower Limestone Shales.

The Millstone Grit is covered for the most part by Secondary Rocks. It consists of shales with subordinate sandstones, and belongs therefore to the type which characterises the South Crop rather than to that of the north side of the Coal Field.

The small tract of Coal Measures which falls within the margin of the map contains representatives of the lower seams, which, however, have been only partly proved and worked to a very limited extent.

The Old Red Sandstone and Carboniferous Rocks were thrown into great east and west folds and subjected to denudation on an enormous scale before the Secondary Rocks were laid down upon them, so that those later strata rest unconformably upon all the earlier formations. The detailed mapping moreover has rendered it possible to recognise many of the features of the pre-Triassic landscape, and to trace step by step the submergence of the old land-surface beneath the Keuper and Rhaetic waters, and its final disappearance in the Liassic sea. The Carboniferous Limestone, then as now, stood out in bold escarpments and formed an island, the margin of which has been traced with much precision.

Some later (post-Liassic) movements have been worked out in detail. They follow generally the lines of the older disturbances, though they rarely agree with them precisely in position.

The map includes a part of what appears to have been the southern margin of the South Wales glacial system. In its north-eastern part large tracts are overspread by a gravelly drift derived from the north and north-east and partly from the immediate neighbourhood. At Pencoed this local drift overlies a stiff boulder-clay containing boulders from a distant source. These erratics, which were first noticed by the late John Storrie, have been microscopically examined by Messrs. Howard and Small, with the result of showing that they can be matched in West Pembrokeshire.

In Chapter XI will be found an account of the economic products of the district. The most important of them was the haematite yielded by the Carboniferous Limestone where that rock was or had been overlain by Keuper. The mines are now all abandoned. In the same chapter the principal sources of water-supply are enumerated.

The map is issued in two editions. On the edition for Solid Geology the glacial deposits are omitted, while on the edition for Superficial Geology, those deposits are shown by colour, as well as the Solid Geology, where not concealed by them.

J. J. H. Teall
Director

Geological Survey Office,
28, Jermyn Street, London

4th November 1904.

ONE

Introduction

GEOGRAPHICAL LOCATION

The Bridgend district (Figure 1) covers most of the mainly agricultural Vale of Glamorgan, with a small part of the South Wales Coalfield in the north and, in the south, a long stretch of the Glamorgan Heritage Coast fringing the Bristol Channel. The largest town is Bridgend, with other substantial settlements at Porthcawl, Pyle, Pencoed, Cowbridge and Llantwit Major.

MAJOR GEOLOGICAL DIVISIONS AND PHYSICAL FEATURES

The main lithological and chronostratigraphical divisions are shown in Figures 1, 2 and 3. The oldest rocks are Devonian sandstones and conglomerates which crop out mainly in the north-east in the core of a major fold, the Cardiff–Cowbridge Anticline. Overlying these is a thick sequence of Dinantian (Lower Carboniferous) limestones, which form escarpments on the northern and southern limbs of the Anticline and crop out westwards along a broad ridge of high ground as far as Porthcawl. Namurian shales and sandstones unconformably overlie the Dinantian limestones in the extreme north of the district, and are in turn overlain conformably by Westphalian Coal Measures.

Mesozoic rocks, including an extensive conglomeratic marginal facies, overlie the Upper Palaeozoic rocks with marked unconformity. Triassic rocks crop out mainly in the north and west, in the east around Pendoylan, and in the vicinity of Cowbridge and Bonvilston. They are overlain by Jurassic limestones and shales, which occupy a large area around Bridgend and form a broad dissected table-land over much of the central and southern parts of the district.

Extensive deposits of Devensian till in the north-east and north-west, have modified the topography and largely obscure the solid formations. Fluvial gravels, ranging from late Devensian to Recent, underlie parts of the broad alluvial

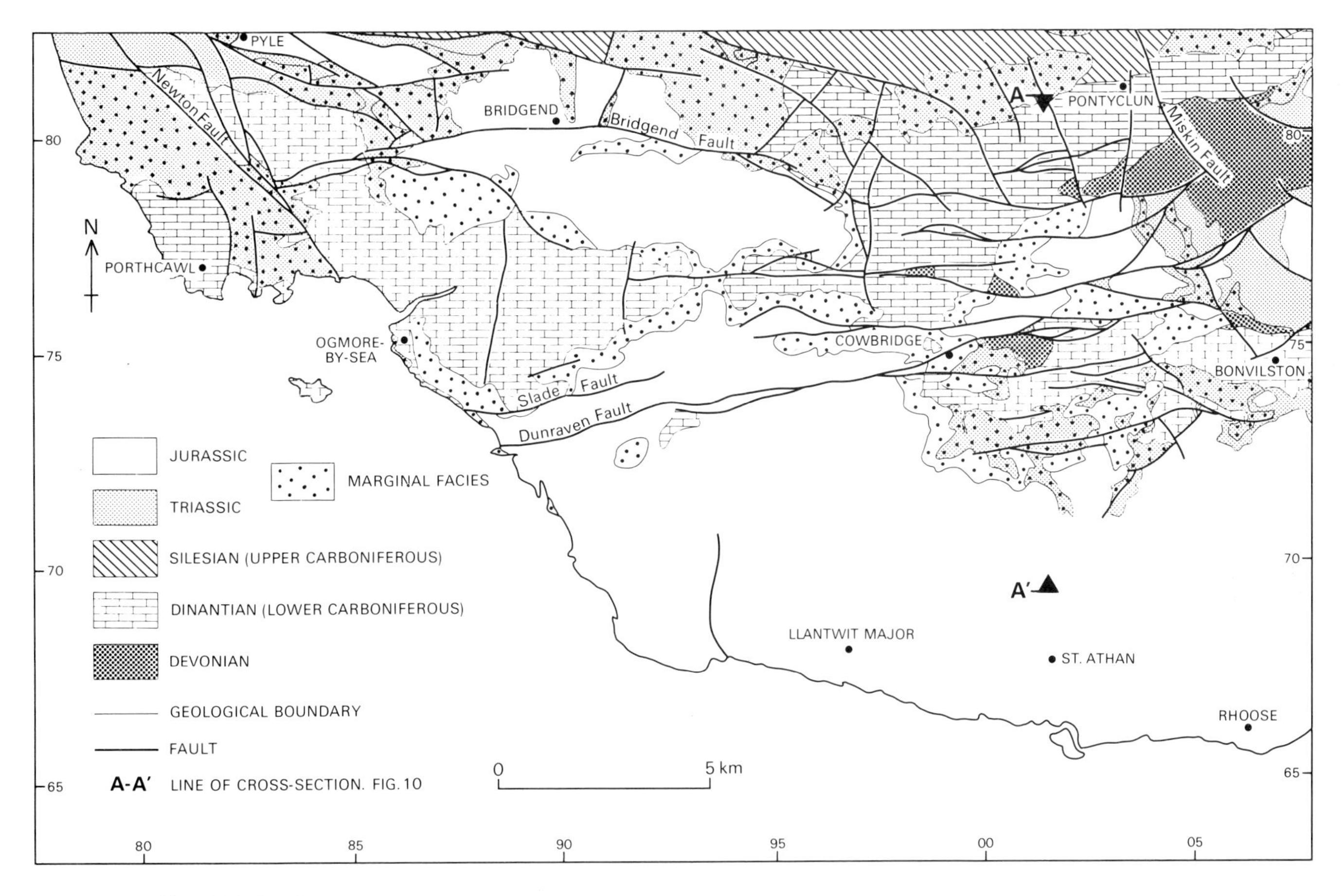

Figure 1 Simplified solid geology map of the Bridgend district

tracts of the Ogmore, Ewenny, Thaw and Ely rivers. Large dune-fields blanket the coastline in the north-west of the district.

TECTONIC FRAMEWORK AND PALAEOGEOGRAPHY

During the Devonian, South Wales formed part of a fluvial coastal plain which was inundated by the Lower Carboniferous marine transgression, establishing carbonate shelf conditions throughout the region. Carbonate environments gave way to prograding deltas and coastal swamps during the Upper Carboniferous. The subsequent folding and uplift, that marked the Variscan Orogeny in South Wales, was followed by a period of extensive erosion prior to the establishment of semiarid fluviatile and lacustrine conditions in the late Triassic. Gradual marine transgression occurred towards the end of the Triassic and during the Lower Jurassic.

The glacial deposits of South Wales are the result of two major Quaternary ice advances, with an intervening interglacial period. The postglacial sediments indicate a change to temperate conditions, which have prevailed to the present day.

Detailed palaeogeographical reconstructions of the Vale of Glamorgan reflect, to a large extent, movements on major basement structures, which were active at intervals during the geological history of the area. These include the Usk Anticline, north-east of Cardiff, and the newly recognised Vale of Glamorgan Axis (formerly the Cardiff–Cowbridge Axis; Waters 1984), a structure coincident with the axial trace of the Cardiff–Cowbridge Anticline.

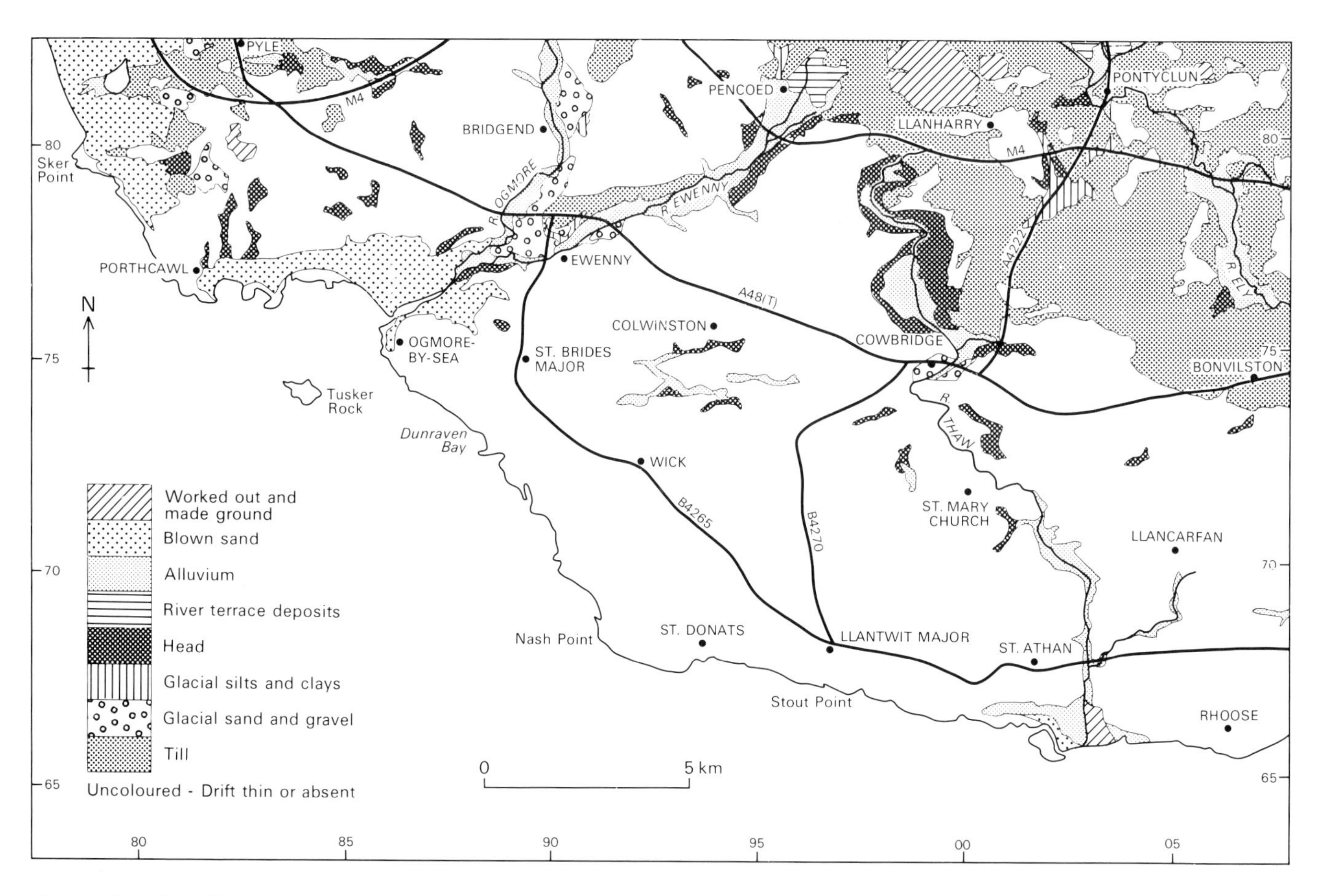

Figure 2 Simplified topographical and drift geology map of the Bridgend district

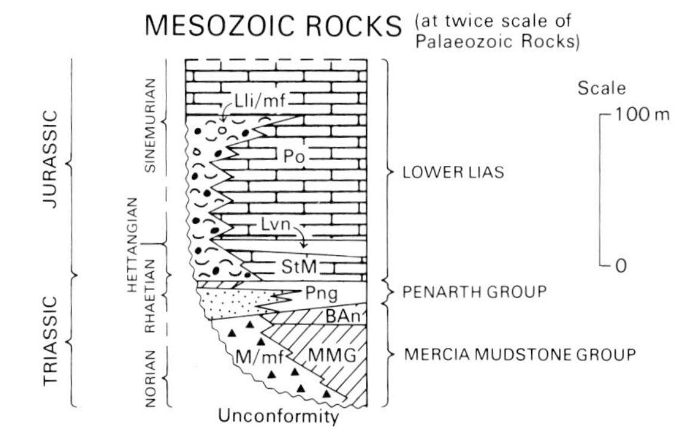

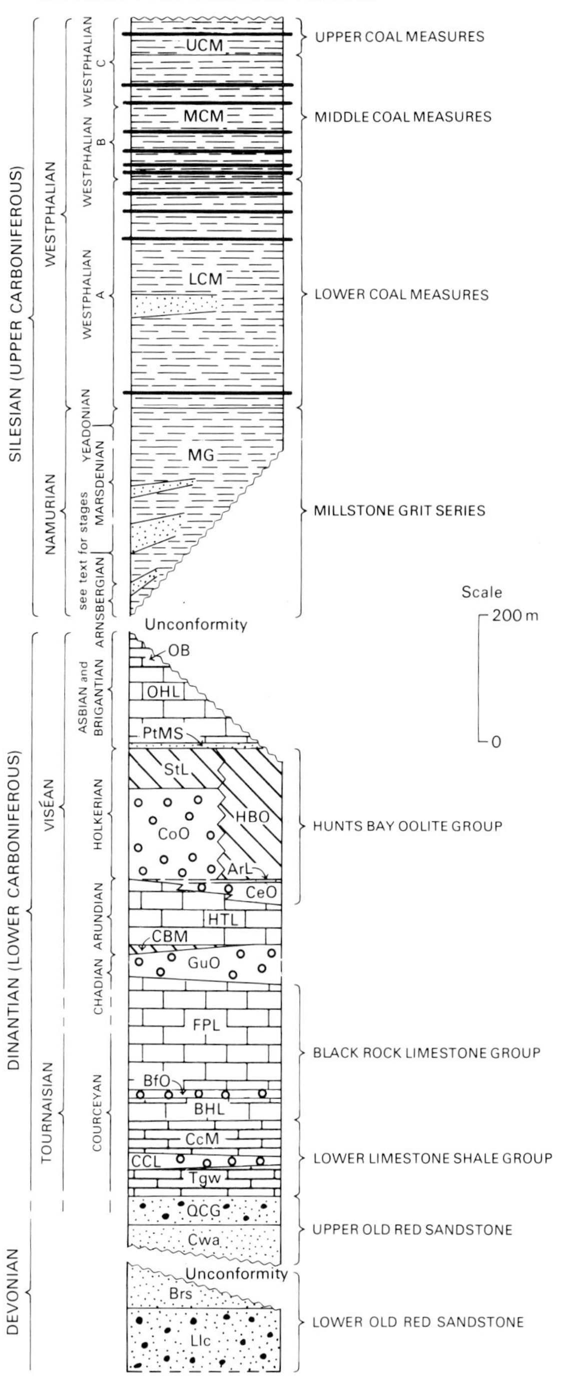

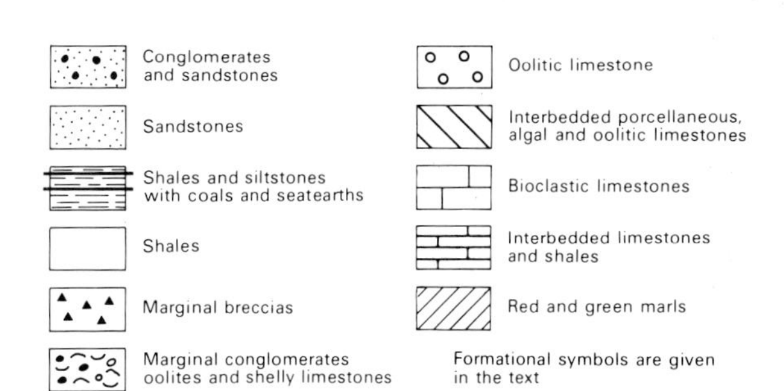

Figure 3 Generalised geological succession of the Bridgend district (only principal lithologies of the formations are indicated)

TWO

Devonian

INTRODUCTION

The Devonian rocks, all of continental Old Red Sandstone facies, are mostly sandstones and conglomerates, deposited in coastal floodplains by rivers flowing dominantly from a northern, semiarid landmass, 'St George's Land'; however, the lowermost Lower Devonian sediments were derived from the south, with a source probably in the region of the Bristol Channel. A major nonsequence between the Upper and Lower Devonian records Middle Devonian uplift and erosion, representing a significant time interval. Close to the top of the Upper Devonian, a conformable passage from fluviatile Old Red Sandstone deposits to marine sequences occurs as a result of a northward marine transgression that progressively drowned the coastal plains.

The Devonian rocks are poorly exposed, but four formations have been identified, largely from criteria established in the adjacent Cardiff district (Waters and Lawrence, 1987). The lowest two, the Llanishen Conglomerate and the overlying Brownstones, are the direct correlatives of Lower Devonian (Lower Old Red Sandstone) rocks of the Newport and Cardiff districts. The overlying Cwrt-yr-ala Formation and Quartz Conglomerate Group comprise the Upper Old Red Sandstone; in the Cardiff district they contain a sparse Upper Devonian fish fauna (Waters and Lawrence, 1987), and range into the Lower Carboniferous (Gayer and others, 1973). There is no marked angular unconformity between the Upper and Lower Devonian, despite the presence of a major stratigraphical break.

LOWER OLD RED SANDSTONE

Llanishen Conglomerate (LlC)

The oldest exposed rocks are the conglomerates, sandstones and siltstones of the Llanishen Conglomerate (Figure 4), which crop out in the core of the Cardiff–Cowbridge Anticline. A thickness of at least 100 m has been recorded, but the base of the formation is nowhere exposed. The conglomerates are typically purplish red, with pebbles and cobbles of volcanic material, sandstone and quartzite, in a coarse sandstone matrix commonly cemented by calcite. They are interbedded with purple- and green-mottled micaceous sandstones and bright red micaceous siltstones and mudstones. The lithologies are generally arranged in a series of fining-upward cycles, commonly with erosional bases. Concretionary limestones within the siltstones and mudstones represent calcrete profiles, which are developed in the upper parts of the cycles. Intraformational conglomerates, dominantly of calcrete clasts, occur at intervals throughout the sequence.

The Llanishen Conglomerate is a proximal alluvial facies (Allen, 1979) and was probably deposited as a series of coalescing alluvial fans. The coarser sediments appear to represent the accumulations of channelised stream systems, and interdigitate laterally with floodplain deposits of siltstones and mudstones in which the calcrete profiles have developed. There is no obvious northerly source for the volcanic pebbles within the conglomerates, but they are similar to Silurian igneous rocks in the Mendips and Bristol areas (Allen, 1975) and it has been suggested that they were derived from the south (Squirrel and Downing, 1969; Allen, 1974).

Locality: Stream section near Clawydd Coch [*ST 0590 7743 to ST 0582 7745*].

Brownstones (Brs)

The Brownstones are very poorly exposed, but crop out on the northern limb of the Cardiff–Cowbridge Anticline, where they attain a thickness of up to 55 m; on the southern limb they are obscured by Mesozoic deposits. A comparison of Brownstones thicknesses between the Cardiff and Bridgend districts demonstrates a pronounced south-westward attenuation. This thinning is attributed to Middle Devonian erosion and subsequent Upper Old Red Sandstone overstep; it has been suggested (Waters and Lawrence, 1987) that it reflects activity on the Vale of Glamorgan Axis. The formation comprises a sequence of drab, purplish red and brown, highly micaceous, cross-bedded, friable sandstones, with subordinate mudstones, siltstones and thin intraformational mudflake conglomerates (Figure 4). Individual sandstone beds, commonly with a conglomeratic lag at their base, are arranged in stacked sequences capped by thin siltstones and mudstones, locally with calcrete profiles; a marked erosion surface into the siltstones and mudstones usually characterises the base of each sandstone sequence.

The Brownstones are generally regarded as deposits of a braided fluvial system, the coarse cross-bedded sandstones and their conglomeratic lags representing the channel fills, and the finer lithologies representing locally preserved overbank deposits (Allen, 1974).

The onset of Brownstones deposition within the Vale of Glamorgan may have been relatively late, for the Llanishen Conglomerate is probably equivalent to the lower part of the Brownstones elsewhere in South Wales and the Borders (Waters and Lawrence, 1987). Therefore, on a regional scale the base of the Brownstones may be markedly diachronous, their southward encroachment into the Vale of Glamorgan resulting from failure of the source area of the southerly derived Llanishen Conglomerate.

Locality: Ditch section, Kennel Grove [*ST 0417 7890*].

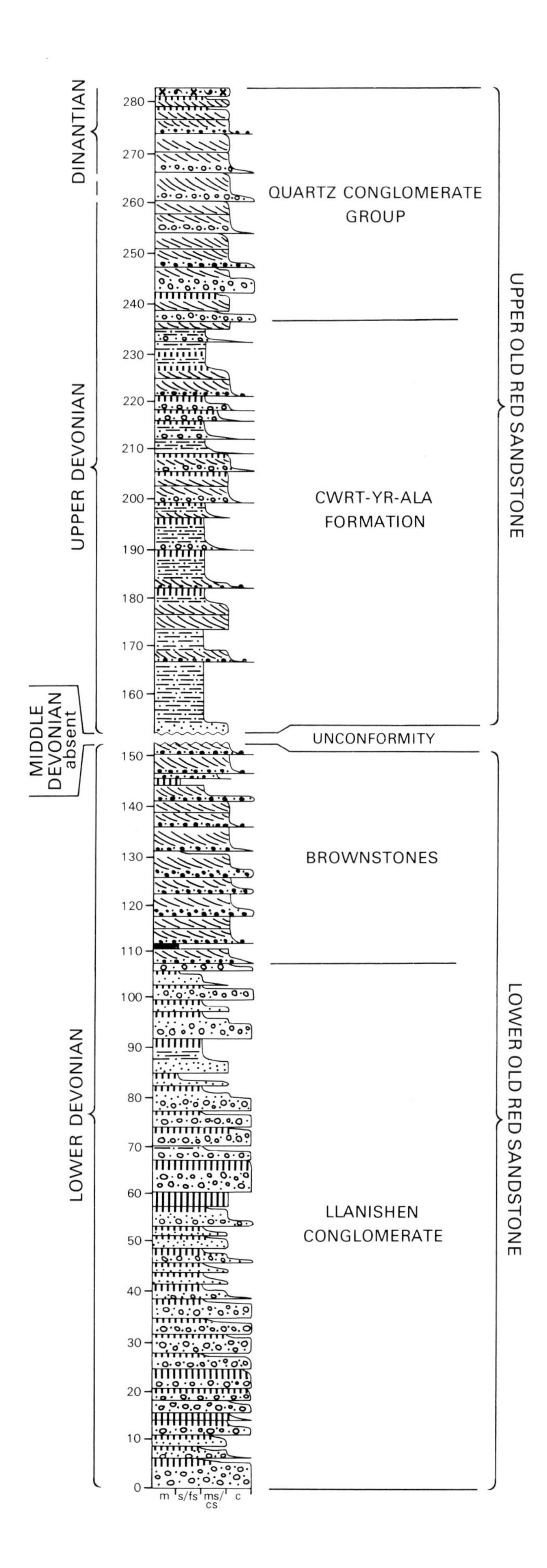

Figure 4 Generalised lithological section for the Devonian of the Vale of Glamorgan

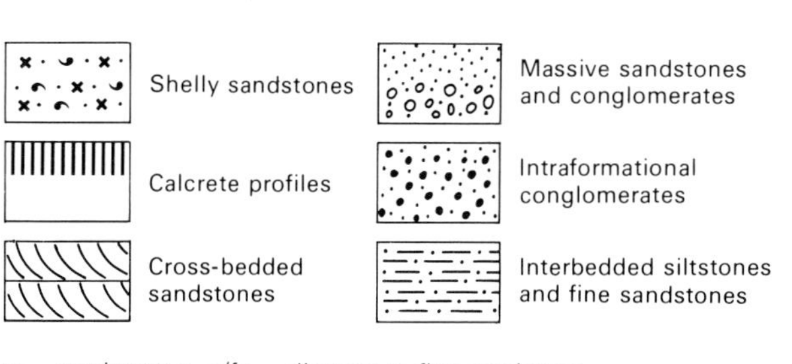

UPPER OLD RED SANDSTONE

Cwrt-yr-ala Formation (Cwa)

The Cwrt-yr-ala Formation (Waters and Lawrence, 1987) crops out on the limbs of the Cardiff–Cowbridge Anticline, but is poorly exposed; small outcrops of undifferentiated Upper Old Red Sandstone (**UORS**), notably near Cowbridge [ST 013 748] and in the Thaw valley [SS 985 767], may include part of the formation. The formation generally comprises reddish brown and pink, planar- and cross-laminated, poorly micaceous, quartzose and locally pebbly sandstones, together with reddish brown siltstones and mudstones containing calcrete profiles (Figure 4). The lithologies are generally arranged in a series of fining-upward fluvial cycles. A typical cycle begins with an erosion surface and a basal conglomeratic lag, overlain by cross-bedded sandstones; these fine upwards rapidly into thick units of thinly interbedded, burrowed sandstones and siltstones, with thin mudstones locally capping each cycle. The cycles are thought to represent floodplain accretion deposits.

The Cwrt-yr-ala Formation sharply overlies the Brownstones and its variation in thickness, from 25 m in the north to over 70 m in the south may, at least partly, be accounted for by synchronous movement on the Vale of Glamorgan Axis.

Locality: Stream section north of Caerwigau Uchaf Farm [ST 0606 7547 to ST 0600 7572].

Quartz Conglomerate 'Group' (QCG)

The Quartz Conglomerate Group comprises 40 to 60 m of pale greenish grey, brown and purple, coarse-grained, pebbly, quartzose sandstones, with subordinate thin siltstones and mudstones (Figure 4). In contrast to the Cwrt-yr-ala Formation, the sandstones are commonly micaceous and exhibit many micaceous partings and laminae. Impersistent thin beds of conglomerate with white quartz pebbles are common in the lower part of the sequence, particularily in the north. The sandstone beds are stacked in thick sequences, capped locally by the siltstones and mudstones and, more rarely, by thin calcrete profiles. Individual sandstone beds are commonly cross-bedded and have erosional bases with thin intraformational conglomeratic lags. These lithologies have been interpreted as the deposits of a low-lying coastal plain or delta. The sandstones appear to represent a number of channel fills, with the siltstones and mudstones representing the overbank deposits; the scarcity of the latter suggests that the channels formed part of a braided fluvial regime (Allen, 1974).

The uppermost 3 to 4 m of the Quartz Conglomerate Group are fine shelly sandstones with siltstones, mudstones and sandy limestones, which contain a marine fauna of brachiopods and crinoids. They probably represent storm-generated marine incursions over the coastal plains as a precursor to the transgression which established marine conditions over the entire district during the Lower Carboniferous. Near Tongwynlais, to the east of the Bridgend district, Gayer and others (1973) have shown that the Devonian–Carboniferous boundary lies within the upper part of the Quartz Conglomerate Group.

Localities: Small crags west of Gadair Wen House [ST 0660 8062]; small quarry near Hensol Lake [ST 0403 7862].

THREE

Dinantian (Lower Carboniferous)

INTRODUCTION

The Carboniferous system in Britain is divided into the Dinantian (or Lower Carboniferous) subsystem and the Silesian (or Upper Carboniferous) subsystem, both of which are present within the district (Figure 3).

The Dinantian of South Wales comprises a southward-thickening prism of alternating bioclastic and oolitic limestones, of predominantly shallow-water origin, which accumulated along, and onlapped onto, the southern flanks of St George's Land, a contemporary upland area of older Caledonian rocks (George, 1958, 1974). Ramsbottom (1973, 1979) has argued that the South Wales succession records a number of sedimentary cycles, separated by nonsequences, which reflect transgressive and regressive movements in sea level that were geographically widespread, and hence eustatic in origin. However, a more detailed appraisal of Dinantian facies variations reveals the controlling effects of major structural axes and lends support to the views of George (1978), who argued for local tectonic control of the distribution of facies and unconformities. It is likely that a dual model, incorporating widely recognised transgressive

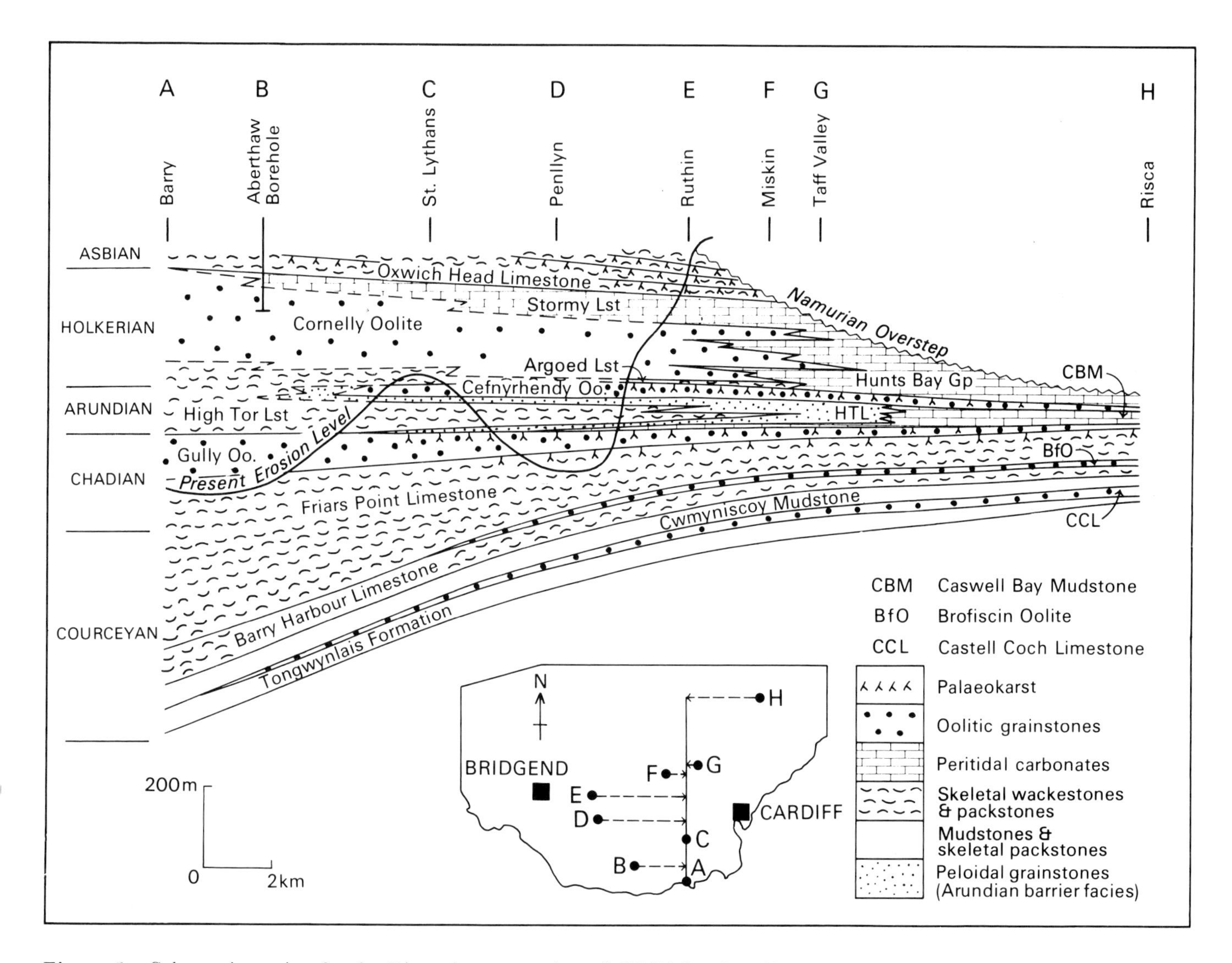

Figure 5 Schematic section for the Dinantian succession of SE Wales. Locality data has been projected onto a common N–S line to illustrate N to S facies and thickness variations. Erosion level refers to areas of Dinantian outcrop and not to areas where Dinantian strata subcrop beneath Mesozoic rocks

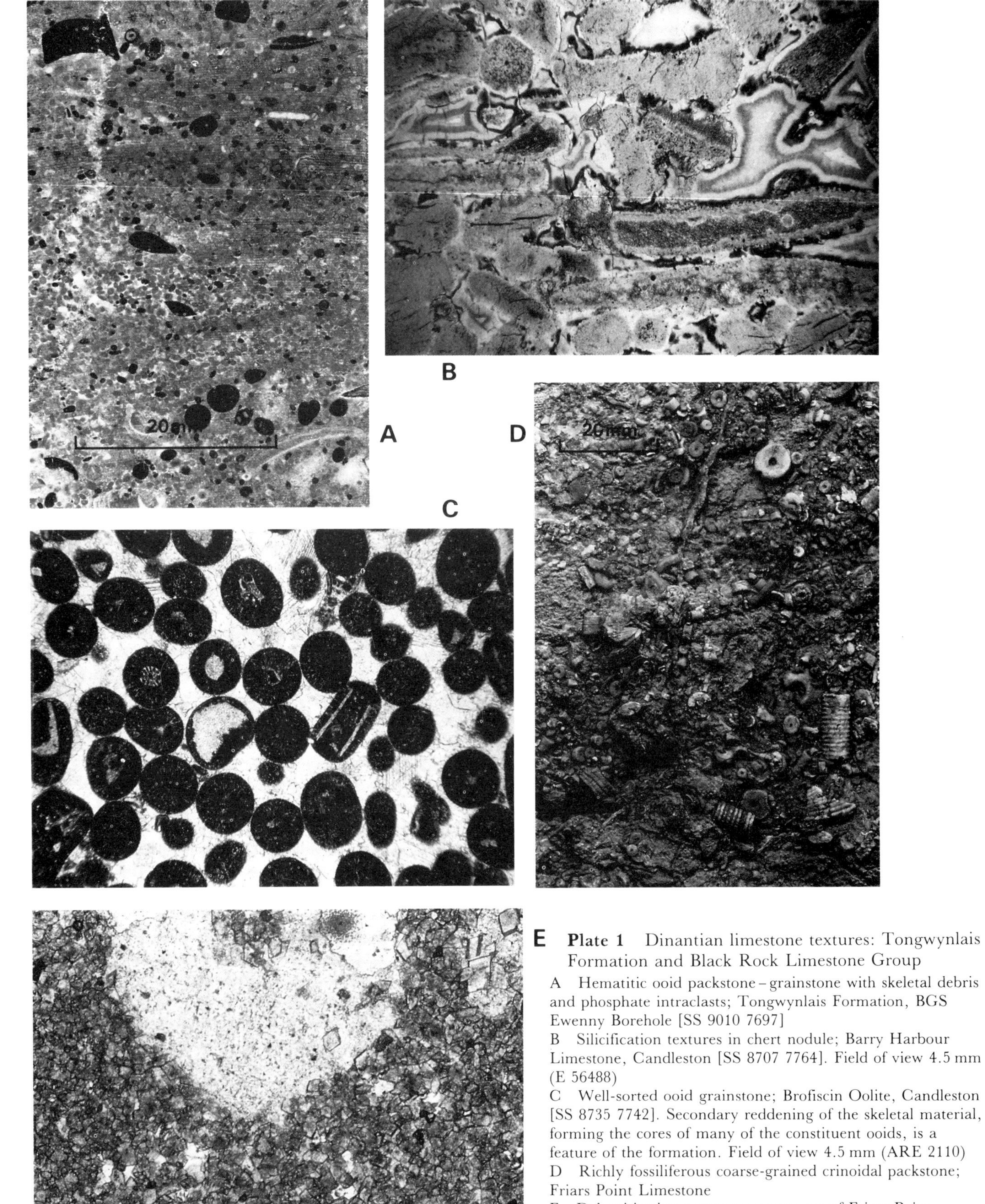

Plate 1 Dinantian limestone textures: Tongwynlais Formation and Black Rock Limestone Group

A Hematitic ooid packstone – grainstone with skeletal debris and phosphate intraclasts; Tongwynlais Formation, BGS Ewenny Borehole [SS 9010 7697]

B Silicification textures in chert nodule; Barry Harbour Limestone, Candleston [SS 8707 7764]. Field of view 4.5 mm (E 56488)

C Well-sorted ooid grainstone; Brofiscin Oolite, Candleston [SS 8735 7742]. Secondary reddening of the skeletal material, forming the cores of many of the constituent ooids, is a feature of the formation. Field of view 4.5 mm (ARE 2110)

D Richly fossiliferous coarse-grained crinoidal packstone; Friars Point Limestone

E Dolomitisation textures; upper part of Friars Point Limestone, River Ogwr [SS 8640 7585]. Small rhombs of dolomite have replaced the limestone matrix and the margins and core of a large crinoid ossicle. Field of view 3.0 mm (E 56462) Note: The field of view of the photomicrograph is its maximum dimension and regressive events superimposed upon an active tectonic framework, can best explain the characteristics of the Dinantian of South Wales.

The base of the Dinantian in the Vale of Glamorgan has been shown to lie within the top few metres of the Old Red Sandstone (Gayer and others, 1973). However, the Dinantian mostly equates with the basal calcareous division of the Carboniferous, the Carboniferous Limestone; a stratigraphical break, of variable magnitude, separates it from the overlying rocks of Silesian age. The lithostratigraphical subdivisions of the Dinantian currently used in the Vale of Glamorgan includes terminology previously used in adjacent areas and newly erected during the course of recent resurveys (Figure 3); the newly erected divisions are indicated in the text. Chronostratigraphical subdivisions are those of George and others (1976).

LOWER LIMESTONE SHALE GROUP

The Lower Limestone Shale Group (Kellaway and Welch 1955) comprises three formations (Waters and Lawrence, 1987); in ascending order these are the Tongwynlais Formation, the Castell Coch Limestone and the Cwmyniscoy Mudstone (Figures 3 and 5). They crop out in the north-east of the district on the limbs of the Cardiff–Cowbridge Anticline, giving rise to well-defined features, although exposures are generally poor. A complete section of the group was provided by the BGS Ewenny Borehole [SS 9010 7697] and it has also been recorded in a borehole on the Bridgend Industrial Estate [SS 9225 7867], where it underlies 196 m of Mesozoic rocks (Appendix 2).

Conodonts and spores, indicative of an early Courceyan age, have been recorded from the Lower Limestone Shale Group (Gayer and others, 1973; Waters and Lawrence, 1987).

Tongwynlais Formation (Tgw)

The Tongwynlais Formation comprises 35 to 45 m of interbedded shales and skeletal packstones with subordinate calcareous sandstones, oolites and hematitic skeletal grainstones (Plate 1a); the latter are equivalent to the alpha-limestones of Dixon and Vaughan (1911). The type section of the formation lies about 6 km to the east of the district (beds 7 to 24 of the sequence described by Gayer and others 1973), where its gradational contact with the underlying Quartz Conglomerate Group is exposed.

The Tongwynlais Formation records the continuation of a marine transgression, initiated during the upper part of the Old Red Sandstone. It marks the establishment of a muddy, sand- and silt-contaminated carbonate shelf with local shoal areas on which oolites were generated. The hematitic grainstones represent high-energy deposits transported away from an environment where primary iron-rich precipitates, probably of chamositic composition, were forming; early alteration to hematite took place during subsequent reworking (Waters and Lawrence, 1987).

Localities: Court Farm [*ST 0108 7529*]; *disused stone pits* [*ST 0672 7522 to ST 0538 8071*].

Castell Coch Limestone (CCL)

The Castell Coch Limestone (the Lower Limestone of Strahan and Cantrill, 1904), resting with sharp, but conformable contact on the underlying Tongwynlais Formation, marks the first regional shallowing event to affect the Dinantian shelf, with the widespread establishment and progradation of high-energy oolite shoals (Burchette, 1981). It ranges from 15 to 25 m in thickness with little systematic variation in the district, and comprises well-bedded, coarse-grained, skeletal ooid grainstones, rich in crinoidal debris, which is locally concentrated into discrete, sharp-based lenses and thin beds of cross-laminated crinoidal packstone. Low-angle and trough cross-bedding is well developed throughout. The formation is commonly reddened by hematite staining.

Localities: Disused quarries between Groesfaen and Miskin [*ST 0539 8075 to ST 0733 8111*]; *Newton Farm* [*SS 9997 7665*]; *roadside quarry, Llanquian-fach* [*ST 0182 7477*].

Cwmyniscoy Mudstone (CcM)

The Cwmyniscoy Mudstone is 45 to 50 m thick over much of the district, but subsequent tectonism has produced local anomalies (e.g. from 30 to 70 m on Stalling Down [ST 015 745]). The formation has a sharp base, and dominantly comprises dark grey, silty, micaceous mudstones with intercalated thin, sandy, argillaceous, shelly limestones. The latter occur as discrete parallel-sided skeletal packstone beds up to 0.3 m thick; individual beds have sharp erosive bases and fine upwards. A variety of trace fossils are commonly preserved on the bases of the beds, but internal bioturbation is usually confined to their upper parts. The intervening mudstone beds become thinner towards the gradational contact with the overlying Black Rock Limestone Group.

The Cwmyniscoy Mudstone marks a return to deeper-water conditions. The lithological and bedding characteristics of the formation are consistent with deposition on a storm-dominated muddy carbonate shelf, the graded limestone beds being the product of individual storm events (cf. Goldring and Bridges, 1973; Dott and Bourgeoise, 1982).

Localities: Road cutting west of Moorlands Farm [*SS 9773 7670*]; *old stone pits at Llansannor* [*SS 9994 7745*]; *stream section south of Pantaquesta Farm* [*ST 0388 7983*].

BLACK ROCK LIMESTONE GROUP

The Black Rock Limestone Group includes all those strata previously assigned to the Black Rock Group (Kellaway and Welch, 1955). These limestones crop out extensively in the east, on both limbs of the Cardiff–Cowbridge Anticline, in the west around Ewenny, and beneath the sand dunes of Merthyr Mawr. The group comprises three formations, the Barry Harbour Limestone, the Brofiscin Oolite and the Friar's Point Limestone. It thickens from about 137 m in the north around Miskin [ST 047 809] to about 270 m in the vicinity of Bonvilston [ST 065 741], and attains a thickness of at least 500 m on the coast at Barry in the adjacent Cardiff district. The generalised facies diagram and cross-section for the Dinantian succession (Figure 5) reveals a marked southward thickening across the Cardiff–Cowbridge Anticline, reflecting syndepositional activity on the Vale of Glamorgan Axis (Waters, 1984).

The Barry Harbour Limestone, Brofiscin Oolite and lower parts of the Friar's Point Limestone all yield conodonts and corals diagnostic of the Courceyan Stage. Upper parts of the Friar's Point Limestone, including dolomites, have been shown to be of Chadian age, containing conodonts of the *Mestognathus beckmanni* biozone and the diagnostic coral *Siphonophyllia cylindrica*.

Barry Harbour Limestone (BHL)

The formation ranges in thickness from 35 m in the north, to 50 m on the southern limb of the Cardiff–Cowbridge Anticline. It generally comprises thinly bedded, dark grey, coarse- to fine-grained, commonly graded, skeletal packstones separated by mudstone partings. The limestone beds vary from parallel-sided to undulate, with erosive bases commonly overlain by coarse crinoidal and shelly lags. Planar, low-angle and hummocky cross-stratification is common, and the disruptive effects of bioturbation are mainly confined to the upper parts of the beds. Some beds are locally highly fossiliferous, being particularly rich in chonetid brachiopods. Silicification (replacement by beekite) of the fossils is common and ubiquitous elongate chert nodules, rich in silicified crinoidal debris, are a distinctive feature (Plate 1b). Intercalated beds of ooid grainstone occur towards the top of the formation in the north-east of the district, but here, dolomitisation has obscured many of the primary depositional textures.

The Barry Harbour Limestone represents a continuation of the storm-influenced depositional regime established during the formation of the Cwmyniscoy Mudstone. Individual limestone beds are similarly interpreted as storm deposits, but the reduced shale content and more varied bedding styles of the Barry Harbour Limestone indicate a higher energy setting on the shelf.

Beds of ooid grainstone towards the top of the Barry Harbour Limestone in the north mark the first influx of sediments derived from an extensive southward-prograding oolite shoal complex that resulted in the widespread deposition of the overlying Brofiscin Oolite.

Localities: Disused quarries between Miskin and Groesfaen [ST 0565 8087, ST 0690 8125 and ST 0730 8125]; disused quarries, east of Rhydhalog [ST 0273 7911] and near Penllyn [SS 977 765].

Brofiscin Oolite (BfO)

The Brofiscin Oolite (the Candleston Oolite of George, 1933) varies in thickness from 13.6 m at Brofiscin Quarry [ST 069 813] to a maximum of 20 m around Penllyn [SS 973 763], but thins to between 8 and 10 m in the Bonvilston area, at Candleston [SS 873 775], and in the BGS Ewenny Borehole (Appendix 2). The formation thins and disappears southwards in the adjacent Cardiff district (Waters and Lawrence, 1987), and a similar attenuation may be anticipated in the Bridgend district (Figure 5).

The Brofiscin Oolite comprises pale to dark grey, well-sorted, ooid grainstones with, locally, well-developed cross-stratification (Plate 1c). The reddened cores of many of the constituent ooids are a distinctive, though not unique feature of the formation.

Localities: Brofiscin Quarry (type section) [ST 069 813]; Hendy Quarry [ST 054 810]; roadside section near Ystradowen [ST 0170 7825]; Candleston Castle [SS 8726 7754]; numerous small sections around Penllyn [SS 975 769].

Friar's Point Limestone (FPL)

This formation sharply overlies the Brofiscin Oolite, but with no evidence of discontinuity. Thickness variations within the Friar's Point Limestone account for much of the overall thickness variation of the Black Rock Limestone Group. The formation ranges from 85 m on the northern limb of the Cardiff–Cowbridge Anticline, to about 210 m at Bonvilston on the southern limb; in the adjacent Cardiff district, on the coast at Barry, over 400 m of the formation are exposed (Waters and Lawrence, 1987).

The Friar's Point Limestone comprises thickly bedded, dark grey to black, foetid, argillaceous skeletal packstones with shaly partings. The effects of extensive bioturbation, enhanced by later pressure solution, imparts an irregular, nodular appearance to many of the beds. The limestones are generally crinoidal and, locally, richly fossiliferous, with a diverse assemblage of corals, brachiopods and bryozoans (Plate 1d).

The upper part of the formation is extensively dolomitised, an alteration recorded throughout South Wales at this stratigraphical level. These upper strata comprise the *Laminosa* Dolomites of previous authors (Dixey and Sibly, 1918; George, 1933). The degree of dolomitisation varies greatly, both vertically and laterally, ranging from scattered rhombs of dolomite in limestone to areas of total alteration to crystalline dolomite (Plate 1e). The dolomitised strata are only mapped within the Friar's Point Limestone where the degree of alteration warrants their distinction; up to 70 m are present in places within the district.

The dolomitisation is entirely a secondary replacement of the bioclastic limestones, with several phases of dolomite formation being recorded (Waters and Lawrence, 1987; Hird and others, 1987). The earlier episodes of dolomitisation were probably intra-Dinantian and related to penecontemporaneous emergence, for example at the top of the Gully Oolite, but subsequent enhancement may have occurred during Triassic or even Tertiary times.

The deposition of the Friar's Point Limestone records a period of renewed transgression, which led to the inundation of the Brofiscin Oolite shoals and re-established deeper-water shelf conditions. Evidence of the storm-induced deposition that characterises the Cwmyniscoy Mudstone and Barry Harbour Limestone is generally lacking, and the extensive bioturbation has largely destroyed any tractional features. In the Bristol district the top of the formation was thought to mark a major nonsequence with the overlying Gully Oolite (Ramsbottom, 1973, 1977). Unfortunately, the widespread dolomitisation of the upper part of the Friar's Point Limestone in the Bridgend district largely obscures the nature of the contact. A discontinuity, possibly related to temporary emergence, is indicated at some localities (e.g. Pant-y-ffynnon Quarry [ST 045 740]) by the presence of a red clay seam resting on an undulating upper surface of the formation. However, whilst a zone of reddening is present around the contact at Ogmore-by-Sea [SS 865 759] and in the BGS Beacons Down Borehole [SS 8874 7521] (Appendix 2), it does not coincide with the top of the Friars Point Limestone, nor is there any marked lithological change across the boundary. Instead, at these localities, the contact with the overlying Gully Oolite appears gradational.

Localities: Brofiscin Quarry [ST 069 813]; Hendy Quarry [ST 054 810]; Pant-y-ffynnon Quarry [ST 045 740]; Longlands Quarry [SS 928 772]; Ewenny Quarry [SS 902 768]; Llantrisant Road cutting [ST 057 811].

GULLY OOLITE (GuO)

The Gully Oolite and its lateral correlatives (George and others, 1976) is an oolitic formation of Chadian age, which occurs throughout South Wales and the Bristol district (the *Caninia* Oolite of previous authors). It comprises thickly bedded, cross-stratified, pale grey, ooid and peloid grainstones with a variable content of skeletal grains (Plate 3a). Discrete beds of skeletal packstone occur throughout, but are more common in the south of the district. Bioturbation is ubiquitous, generally as vertical burrows which disrupt cross-lamination and cause digitate contacts between oolitic and non-oolitic beds. Superficial and compound ooids are common (Plate 3a), the latter attaining pisolitic dimensions. Cross-stratification occurs on several scales, in sets ranging from 2 to 3 cm up to 4 m in thickness; scour-and-fill structures abound.

In the west of the district the formation includes a distinctive basal unit, the **Ogwr Member** (*new name*), which is not distinguished on the 1:50 000 map. It comprises up to 7 m of coarse-grained, flaggy, crinoidal packstones and fine-grained, low-angle cross-bedded skeletal packstones. The chonetid brachiopods *Delepinea notata* and *Megachonetes magma*, suggesting a Chadian age, occur in abundance throughout the Ogwr Member, which is comparable, both in lithology and stratigraphical position, to the suboolite bed of the Bristol district (Vaughan, 1905; Kellaway and Welch, 1955).

The Gully Oolite shows a marked southward increase in thickness across the hinge of the Cardiff–Cowbridge Anticline (Figure 5), from 30 m in the north-east to about 70 m around Bonvilston. This rapid thickness variation records the continued operation of the syndepositional hinge-line of the Vale of Glamorgan Axis.

The Gully Oolite was deposited as part of a regional development of oolite shoals, generated in response to Chadian marine regression. Intercalated skeletal packstones in the Ogmore area indicate a local transition into an offshore bioclastic facies. The top of the Gully Oolite marks an important and widely recognised nonsequence often referred to as the mid-Avonian unconformity (Dixon and Vaughan, 1912). The upper surface of the formation is irregular and pitted and locally reduced to an ill-defined rubbly zone. The detached blocks of oolite and the underlying in-situ limestones exhibit textures and fabrics diagnostic of calcretes; rhizoliths (alteration halos developed around roots; Klappa, 1980) are locally well displayed (Plate 3b). Mottled grey-green clays overlie this horizon and also occur as matrix material within the rubbly zones. These phenomena are interpreted as a fossil soil profile overlying a palaeokarst surface (Riding and Wright, 1981).

Emergent effects at the top of the Gully Oolite have only been observed on the northern limb of the Cardiff–Cowbridge Anticline (but see below).

Localities: Hendy Quarry [ST 054 810]; [Forest Wood Quarry]ST 015 797; Dutchy Quarry [SS 906 757]; Lancaster Quarry [SS 908 755]; Pontalun Quarry [SS 897 765]; Pant Quarry [SS 897 760]; River Ogmore [SS 8645 7588]; Llantrisant Road cutting [ST 057 811].

CASWELL BAY MUDSTONE (CBM)

The Arundian sea-level rise and the development of a temporary barrier in the south of the district established a lagoonal/peritidal setting for the deposition of the Caswell Bay Mudstone. This thin, though distinctive formation (George and others, 1976; Riding and Wright, 1981) corresponds to the *Modiola* phase at the base of the C_2 zone (Dixey and Sibly, 1918). It comprises interbedded calcite mudstones, shales, argillaceous skeletal wackestones and packstones. Cryptalgal lamination and spar-filled bird's-eye structures are well displayed in some beds, and bioturbation mottling is widely developed. Parallel-sided beds of skeletal, peloid grainstone up to 1.3 m thick, with sharp erosive bases, are intercalated in the upper part of the formation (Plate 2).

In the north of the district the Caswell Bay Mudstone varies from 3.5 to 14.7 m in thickness, largely reflecting broad irregularities in the underlying palaeokarstic surface. The formation is not observed on the southern limb of the Cardiff–Cowbridge Anticline, although its occurrence in the BGS St Lythans Borehole (Waters and Lawrence, 1987), immediately east of the district, suggests that it may be present in an attenuated form, in the Cowbridge to Bonvilston area. The formation is certainly absent from the southern and western parts of the district, where the succeeding High Tor Limestone rests with a sharp, planar, and possibly erosive, basal contact on unmodified Gully Oolite. The formation of an east–west barrier complex in the south, which restricted the Caswell Bay Mudstone to the northern parts of the district, indicates the continuing control of the Vale of Glamorgan Axis on shelf bathymetry (Figure 5). The grainstone beds towards the top of the Caswell Bay Mudstone and at the base of the succeeding High Tor Limestone record the transgressive passage of this barrier, driven northwards as the Arundian sea-level continued to rise (Plate 2).

Localities: Forest Wood Quarry [ST 015 797]; Hendy Quarry [ST 054 810].

Plate 2
Caswell Bay Mudstone (CBM) overlying Gully Oolite (GuO) and overlain by High Tor Limestone (HTL); Forest Wood Quarry [SS 0165 7970]

The High Tor Limestone is dominantly of peloidal packstone–grainstone facies. The prominent bed in the upper part of the Caswell Bay Mudstone is a peloidal grainstone resulting from washover of the Arundian barrier complex into the CBM lagoon (A 14293)

HIGH TOR LIMESTONE (HTL)

The High Tor Limestone (George and others, 1976) crops out extensively on the northern limb of the Cardiff–Cowbridge Anticline, in the vicinity of Ogmore-by-Sea [SS 865 752] and St Brides Major [SS 896 751], and on Tusker Rock [SS 841 742]. The formation exhibits significant variations in thickness, sequence and facies across the district (Figure 5; Plate 3c and d)). The complete High Tor Limestone sequence of the Ogmore and St Brides area was established in the BGS Beacons Down Borehole [SS 8874 7521] (Appendix 2), where it is 133 m thick and comprises three parts. The basal 74 m consists predominantly of well-bedded, skeletal packstones and wackestones with intercalated thin shales (Plate 4). The limestones vary from fine grained to coarsely crinoidal and contain a rich and diverse Arundian fossil assemblage (Plate 3c), which includes the characteristic brachiopod *Delepinia destinezi*, the solitary corals *Siphonophyllia garwoodi* and *Palaeosmilia murchisoni*, and a diagnostic microfauna, with the foraminifera *Eoparastaffella simplex* and *Mediocris breviscula* and the dasycladacean alga *Koninckopora inflata*. These beds pass upwards into 44 m of thickly bedded, locally cross-stratified, skeletal peloid packstone–grainstones with scattered ooids (Plate 3d). The uppermost 15 m of the formation has a sharp lower contact and comprises a basal calcareous mudstone bed overlain by a sequence of skeletal packstones. This upper division contains a diagnostic Holkerian microfauna and is correlated with the Argoed Limestone Member in the north of the district (Figure 5), its basal contact marking the local Arundian–Holkerian stage boundary.

The total thickness of strata assigned to the High Tor Limestone in the north-east is only about 60 m. At Forest Wood Quarry, a basal division of coarse, crinoidal packstone–grainstones up to 3 m thick, rests with erosive contact on the Caswell Bay Mudstone. It is succeeded by up to 40 m of well-bedded skeletal packstones, locally with peloidal and oolitic beds and low-angle cross-bedding (Plate 2); in contrast to the sequence in the St Brides area, the shale content is much reduced. The packstones give way to an upper unit, 15 to 20 m thick, of hummocky cross-stratified, skeletal, peloidal packstone–grainstones, which become increasingly oolitic towards the gradational contact with the overlying Cefnyrhendy Oolite Member, the basal division of the succeeding Hunts Bay Oolite Group (Figure 3).

The barrier complex, which restricted the peritidal sequences of the Caswell Bay Mudstone to the north of the district during the early Arundian, migrated progressively northwards in response to continuing transgression (Figure 5). The fossiliferous skeletal packstones and wackestones of the High Tor Limestone, overlying the Gully Oolite in the south-west, demonstrate the early establishment of an offshore, open marine shelf environment in this area; the absence of tractional sedimentary structures in these strata points to the obliterative effects of subsequent bioturbation. With the migration of the barrier complex, open marine conditions were subsequently established in the north of the district, but the character of the lowermost High Tor Limestone in this area, with little shale, oolitic in parts and with cross-stratification locally preserved, confirms that it remained in a shallower, near-shore setting than that in which its southern counterparts were deposited.

Upper parts of the High Tor Limestone in the north, record the subsequent progradation back into the district of a high-energy shoreface environment, in which the hummocky cross-stratified packstone–grainstones were deposited. This

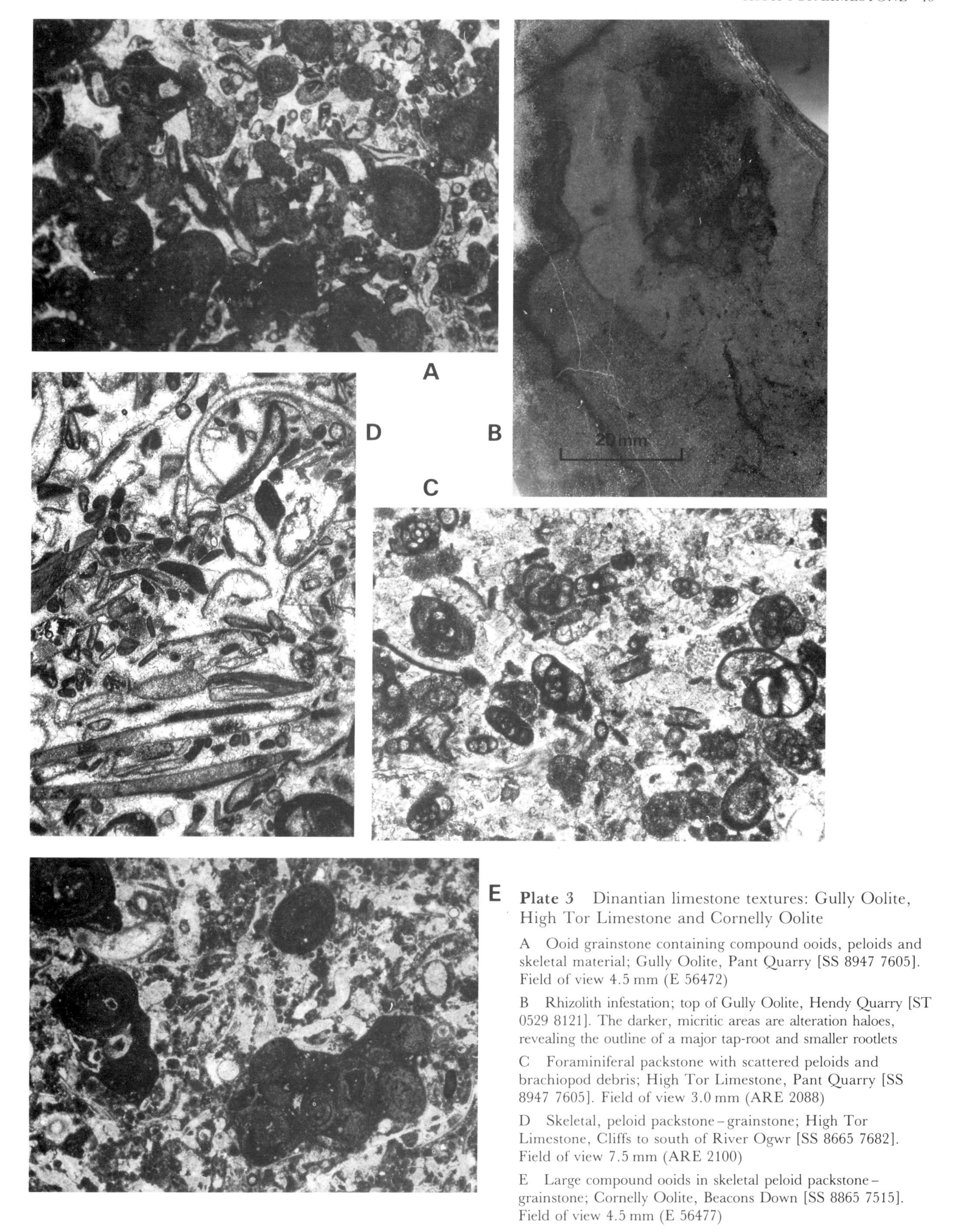

Plate 3 Dinantian limestone textures: Gully Oolite, High Tor Limestone and Cornelly Oolite

A Ooid grainstone containing compound ooids, peloids and skeletal material; Gully Oolite, Pant Quarry [SS 8947 7605]. Field of view 4.5 mm (E 56472)

B Rhizolith infestation; top of Gully Oolite, Hendy Quarry [ST 0529 8121]. The darker, micritic areas are alteration haloes, revealing the outline of a major tap-root and smaller rootlets

C Foraminiferal packstone with scattered peloids and brachiopod debris; High Tor Limestone, Pant Quarry [SS 8947 7605]. Field of view 3.0 mm (ARE 2088)

D Skeletal, peloid packstone – grainstone; High Tor Limestone, Cliffs to south of River Ogwr [SS 8665 7682]. Field of view 7.5 mm (ARE 2100)

E Large compound ooids in skeletal peloid packstone – grainstone; Cornelly Oolite, Beacons Down [SS 8865 7515]. Field of view 4.5 mm (E 56477)

Plate 4 Well-bedded, argillaceous skeletal packstone–wackestones with shaly partings; High Tor Limestone, Ogmore-by-Sea [SS 8660 7444] (A 14287)

late-Arundian shoaling event culminated in the deposition of the overlying Cefnyrhendy Oolite in the north (Figure 5). In the distal south-western areas, this capping unit of ooid grainstone was not developed and, instead, the shallowing episode is represented by the sequence of thickly bedded packstone–grainstones with scattered ooids.

Localities: Forest Wood Quarry [ST 015 797]; Argoed Isha Quarry [SS 993 790]; Pant St Brides [SS 894 756]; the area south of Mount Pleasant Farm [SS 834 794]; Ogmore-by-Sea [SS 865 745].

HUNTS BAY OOLITE GROUP (HBO)

The Hunts Bay Oolite Group crops out extensively within the district and is mostly Holkerian in age, containing the characteristic brachiopods *Davidsonia carbonaria*, *Linoprotonia corrugatohemisphericus* and *Composita ficoides*, the last being particularly abundant in the upper parts of the group. In the north-east however, strata of Arundian age are also included in the group (Figure 3), and the thickness increases accordingly from about 200 m in the west, to a maximum of 245 m in the east; the thickness diminishes rapidly, however, to the east of the Miskin Fault as a result of Namurian overstep and unconformity (Figure 5).

It has long been recognised that the Hunts Bay Oolite Group is divisible into two broad lithological units (Dixon and Vaughan, 1911; Kellaway and Welch, 1955); these are now named the Cornelly Oolite Formation and the overlying Stormy Limestone Formation. This subdivision holds as far east as Pontyclun [ST 038 809], beyond which the two divisions 'pass insensibly into one another' (Dixey and Sibly, 1918) as lithologies characteristic of the Stormy Limestone Formation occur at progressively lower levels within the oolite formation (Figure 5).

Cornelly Oolite Formation (CoO)

The Cornelly Oolite Formation (*new name*) crops out around South Cornelly [SS 825 796] and Newton Down [SS 840 796], in the vicinity of St Brides Major [SS 895 749], and on the northern limb of the Cardiff–Cowbridge Anticline between Ruthin [SS 970 797] and Pontyclun. The bulk of the formation comprises thickly bedded, cross-stratified, ooid grainstones, commonly with large compound ooids that attain pisolitic dimensions (Plate 3e). Intraclasts of reworked, early cemented oolite form conglomeratic lags that line scour-and-fill structures and occur within the toe-sets of cross-bedding. Thin packages of skeletal packstone occur locally throughout the formation, but are generally not correlatable. Between Ruthin and Pontyclun, beds of calcite mudstone and oncolitic peloid grainstone occur within the Cornelly Oolite; these lithologies, which typify the overlying Stormy Limestone Formation, herald the progressive breakdown of the two-fold division of the Hunts Bay Group as it is traced eastwards.

In the west, on Newton Down, the formation attains a thickness of 135 m, increasing southwards to 150 m around St Brides Major. In the east, it varies from 150 to 180 m and includes two additional divisions at the base, the **Cefnyrhendy Oolite Member** (**CeO**; *new name*) and the **Argoed Limestone Member** (**ArL**; *new name*). The former, comprising well sorted, ooid grainstones with a diagnostic Arundian assemblage of foraminifera and algae, gradationally overlies the High Tor Limestone and displays a sharp top which, locally, may represent a palaeokarstic surface. It ranges in thickness from 13 m around Groesfaen [ST 064 815] to 40 m in the Ruthin area. The overall thickness variation of the Cornelly Oolite Formation along its eastern crop is partly attributable to the appearance of this basal oolite member. The overlying Argoed Limestone, up to 10 m thick, comprises a distinctive basal bed of calcareous mudstone, overlain by thin- to medium-bedded, fine to coarse, crinoidal skeletal packstones, locally rich in *Siphonodendron martini* and yielding the Holkerian foraminifera *Archaediscus* at *concavus* stage and *Koskinotextularia* (Waters and Lawrence, 1987). These basal members are correlated with the upper part of the High Tor Limestone in the south-west of the district (Figure 5).

Deposition of the progradational Cefnyrhendy Oolite in the north-east records the continuation of the Arundian shoaling event, which culminated, locally, in emergence. The succeeding Argoed Limestone records the brief but widespread re-introduction of offshore conditions in response to a basal Holkerian marine transgression. Colonies of lithostrotionid corals flourished during this period prior to the re-establishment of a more extensive Cornelly Oolite shoal complex, which blanketed much of the district. The thickness of this oolitic division suggests that deposition kept pace with a slowly, but steadily, rising sea-level, so that a shallow, ooid-generating environment was sustained.

Localities: Cornelly Quarry (type section) [SS 835 800]; Grove Quarry [SS 822 797]; Stormy Down Quarry [SS 842 804]; Ruthin and Garwa Farm Quarries [SS 975 795]; Argoed Isha Quarry (type section of Argoed Limestone) [SS 993 790]; Cefnyrhendy Quarry (type section of Cefnyrhendy Oolite) [ST 0677 8129].

Stormy Limestone Formation (SrL)

The Stormy Limestone (*new name*) corresponds to the previously termed 'Modiola' or 'Lagoon' phase at the top of the S_2 subzone (Dixon and Vaughan, 1912; Dixey and Sibly, 1918; George, 1933). It maintains a fairly constant thickness of 60 to 65 m across the north of the district, but in the south it thins to 55 m in an inlier at St Andrews Minor [SS 928 733]. The Stormy Limestone is a heterolithic formation, mainly consisting of interbedded porcellaneous calcite mudstones, cocquinoid limestones and packstone–grainstones (Plate 6a–f). The calcite mudstones occur as thin beds, commonly with stromatolitic lamination and birds-eye structures. The coquinoid limestones are composed predominantly of articulated and disarticulated valves of *Composita ficoides*. The units of packstone–grainstone contain variable proportions of skeletal grains, ooids, peloids and intraclasts (Plate 6e and f). A high proportion of ooid grainstone is maintained throughout the formation. The abundance of oncolites is diagnostic of the Stormy Limestone; these occur in discrete beds of oncolitic grainstone, but also scattered throughout the other lithologies (Plate 6a, b and e). Beds rich in gastropods and the sclerosponge *Chaetetes* also occur. The microfossils yielded by the Stormy Limestone, including the foraminifera *Klubonibelia* sp., *Eostaffella parastruvei*, *Septabrunsiina* sp. and the alga *Koninckopora inflata* in abundance, form a distinctive and specialised assemblage, indicative of restricted shallow-water conditions (Strank, 1982; Somerville and Strank, 1984).

Thrombolitic algal bioherms, up to 2 m high and 3 to 4 m in diameter, occur within the formation in Garwa Farm Quarry [SS 980 797] (Plate 5). The surrounding beds drape over these massive lenticular and loaf-shaped structures and are loaded down beneath them. Internally, the bioherms are composed of ramifying clots, tufts, stringers and lenses of microcrystalline and cryptocrystalline calcite within a fine skeletal wackestone matrix; vermetid gastropods are also present (cf. Burchett and Riding, 1977).

The base of the Stormy Limestone is well defined in the west of the district, where it is taken at the lowest recorded calcite mudstone bed within the local Hunts Bay Group succession. This bed is overlain by a thin (up to 7 cm) red clay seam, in which well developed desiccation polygons are preserved (Plate 7). In the extreme north-east, however, the occurrence of lithologies characteristic of the Stormy Limestone at progressively lower levels within the Holkerian succession, makes the subdivision of the Hunts Bay Group impractical (Dixey and Sibly, 1918) except for the delineation of the basal Cefnyrhendy Oolite and Argoed Limestone Members (Figure 5). The top of the Stormy Limestone Formation, where it is overlain by the Oxwich Head Limestone, is marked by stratigraphical discontinuity, displayed as an irregular, hummocky palaeokarstic surface, overlain by a mottled red and grey clay representing a palaeosol.

The heterolithic nature of the Stormy Limestone formation is indicative of deposition in a variety of lagoonal/peritidal back-barrier environments which developed behind the Cornelly Oolite shoal complex, including inter- and supratidal flat, tidal channel and delta, washover fan and deep lagoon (Ginsberg, 1975; Scholle and others, 1983). In the north-east of the district, where these lithofacies occur throughout the Hunts Bay Group, a lagoonal setting was evidently maintained throughout the Holkerian. Westwards, and possibly southwards, progradation of the Cornelly Oolite allowed lagoonal conditions to become established over a progressively wider area, culminating in deposition of the Stormy Limestone over most of the district.

Marine regression terminated Holkerian deposition and led to the formation of the palaeokarstic surface at the top of the Hunts Bay Group.

Localities: Stormy Down Quarry (type section) [SS 842 804]; Cornelly Quarry [SS 835 800]; Grove Quarry [SS 822 797]; Garwa Farm Quarry [SS 980 797]; Pant Mawr Quarry [SS 828 802]; Stormy West Quarry [SS 846 806].

OXWICH HEAD LIMESTONE (OHL)

The Oxwich Head Limestone (George and others, 1976) crops out extensively in the west, around Porthcawl [SS 813 770] and South Cornelly, and on the northern limb of the Cardiff–Cowbridge Anticline between Pencoed and Pontyclun. The outcrop thins dramatically eastwards from Pencoed, in response to the sub-Namurian unconformity, and has been removed completely to the east of the Miskin Fault. A further small outcrop, representing the lowest 50 m of the formation, occurs in an inlier at St Andrews Major [SS 927 735]. In the extreme west of the district the formation is about 130 m thick, although faulting and poor exposure

Plate 5 Algal bioherms (4 m across) within the Dinantian Stormy Limestone; Garwa Farm Quarry [SS 980 797] (A 14308)

preclude an accurate assessment; 68 m of Oxwich Head Limestone were proved in the BGS Porthcawl Borehole [SS 8112 7773] (Appendix 2).

The basal member of the Oxwich Head Limestone is the **Pant Mawr Sandstone (PtMS**; *new name*). This unit is up to 4 m thick in the west, but includes a lower 1.5 m of partially dolomitised skeletal packstones with only scattered sand grains. The upper half comprises thin, flaggy beds of fine-grained calcareous sandstone (Plate 9a) with low-angle cross-bedding and wave-ripple cross-lamination, locally disturbed by subsequent bioturbation. The contact of the sandstone with the overlying limestones is generally sharp, although the lowest metre or so of the latter is also commonly sandy. In the eastern outcrops the basal packstone phase is absent, and the member, reduced to 1.5 m in thickness, is composed wholly of calcareous sandstone. Honeycomb weathering is a distinctive feature of the deposit. The Pant Mawr Sandstone is tentatively correlated with the Honeycombed Sandstone of the North Crop (George and others, 1976); the Middle Cromhall Sandstone of the Bristol District also occupies a similar stratigraphical position (Kellaway and Welch, 1955).

The main part of the Oxwich Head Limestone comprises thickly bedded, fine- to coarse-grained, recrystallised skeletal packstones (Plate 9c), with distinctive light to dark grey mottling and pseudobrecciation (Dixon and Vaughan, 1911). Oolitic, peloidal and coquinoid grainstones (Plate 9b) occur locally towards the base and compare with the Penderyn Oolite of the North Crop (George and others, 1976). Units, up to 5 m thick, of dark grey, irregularly bedded skeletal packstones with shaly partings are developed at intervals and are well exposed in coastal sections at Porthcawl.

The formation is locally highly fossiliferous, with abundant solitary corals and productid brachiopods. Lower beds have yielded ambiguous foraminiferal assemblages containing both Holkerian and Asbian forms. The former are thought to have been reworked from the underlying Hunts Bay Group, since the occurrence in the basal bed at Pant Mawr Quarry of *Bibradya inflata*, a diagnostic Asbian form, confirms the position of the stage boundary at the level of the underlying palaeokarst. Higher parts of the formation, in Gaen Quarry [SS 824 804] and coastal sections at Porthcawl, contain typical Asbian foraminiferal assemblages including *Archaediscus karreri*, *Endothyra tantala*, *Gigasbia gigas* and double-walled *Palaeotextularia*. Diagnostic macrofauna from the formation include the corals *Dibunophyllum bourtonense*, *Palaeosmilia murchisoni* and *Lithostrotion junceum*. The upper parts of the Oxwich Head Limestone probably range into the Brigantian stage, as in the Gower Peninsula (George and others, 1976).

Plate 6 Dinantian limestone textures: Stormy Limestone

A Oncolites with spar-filled fenestrae and skeletal nuclei; Stormy Limestone, Tythegston Quarry [SS 8510 7886]. Field of view 7.5 mm (E 56495)

B Compacted oncolitic mudstone; Stormy Limestone, BGS Porthcawl Borehole [SS 8112 7773]

C Cryptalgal lamination; Stormy Limestone, BGS Porthcawl Borehole [SS 8112 7773]

D Intraclast, peloidal, skeletal packstone – grainstone with spar-filled shelter cavities beneath disarticulated *Composita* valves. A small colony of *Chaetetes* occurs in the lower left of the photograph. Stormy Limestone, BGS Porthcawl Borehole [SS 8112 7773]

E Intraclast, oncoid rudstone; Stormy Limestone, BGS Porthcawl Borehole [SS 8112 7773]

F Peloidal, skeletal grainstone; Stormy Limestone, Pant Mawr Quarry [SS 8275 8033]. Field of view 4.5 mm (ARE 2050)

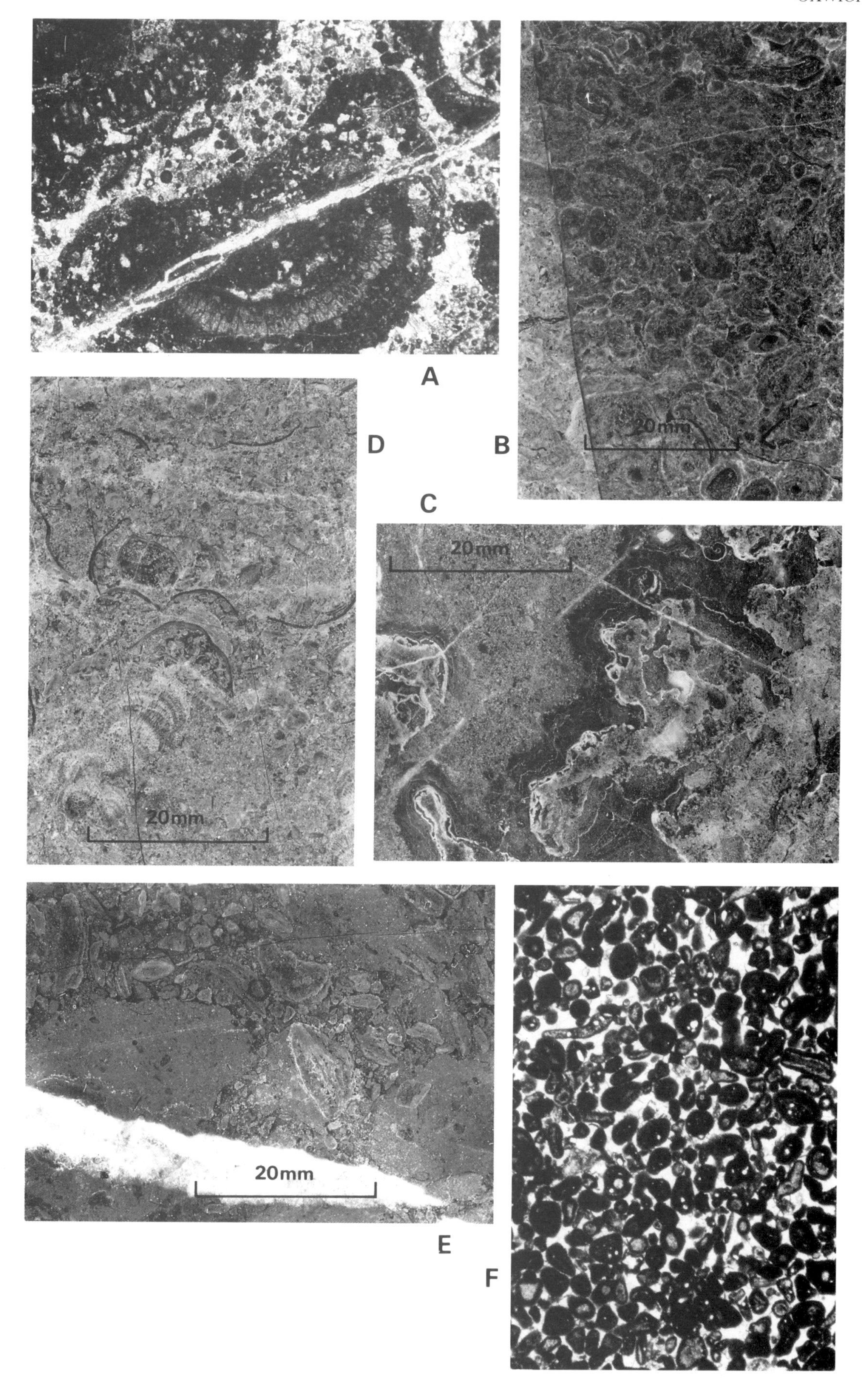
A
B
20mm
D
C
20mm
20mm
E
20mm
F

Plate 7 Cast of desiccation cracks in the Dinantian Stormy Limestone; Cornelly Quarry [SS 835 800] (Photography by J R Davies)

Plate 8 Hummocky palaeokarstic surface (foreground) within Oxwich Head Limestone; Locks Common, Porthcawl [SS 8075 7700]. The overlying bed to the right pinches out against the irregularities in the palaeokarstic surface (A 14326)

Conspicuous irregular, hummocky and pitted surfaces, commonly overlain by mottled red and grey clay seams, punctuate the Oxwich Head Limestone at intervals of between 7.5 and 21 m. These features are interpreted as palaeokarstic surfaces and palaeosols, the products of penecontemporaneous subaerial exposure (Plate 8). The underlying host limestones exhibit pedogenic alteration effects including infestation by rhizoliths. Such surfaces are now widely recognised in strata of this age throughout the country (e.g. Walkden, 1974), but were first recorded by Tiddeman (in Strahan and Cantrill, 1904, p.14) from localities in the Bridgend district. The palaeosols are generally less than 0.25 m thick, but locally fill deep hollows in the palaeokarstic surfaces up to 6 m deep and 40 m wide (Thomas, 1953). The surfaces impart a minor cyclicity to the Oxwich Head Limestone, a feature typical of Asbian and Brigantian strata throughout Britain (Ramsbottom, 1973).

The marine regression at the end of the Holkerian probably resulted in terrigenous quartz sand being carried onto the emergent carbonate shelf to undergo basal Asbian marine reworking and redeposition as the transgressive Pant Mawr Sandstone. The remaining parts of the Oxwich Head Limestone record cyclical deposition on a mature carbonate platform, where minor fluctuations in sea-level, eustatic or tectonically induced, resulted in either the inundation or exposure of extensive areas of the shelf. The mottled and pseudobrecciated packstones, which form the body of the cycles, were formed during periods of high sea-level. They record the blanket deposition across the shelf of thoroughly bioturbated carbonate muddy sands, deposits typical of many modern carbonate platforms (Ginsburg and James, 1974). During periods of lower sea-level the numerous and distinctive palaeokarstic surfaces were formed. Red and grey mottled clay palaeosols accumulated on these surfaces and were colonised by vegetation. Rootlets penetrated and infested the host limestones to produce the observed rhizoliths.

Localities: *Gaen Quarry* [*SS 824 804*]; *Grove Quarry* [*SS 822 797*]; *Garwa Farm Quarry* [*SS 975 795*]; *coastal exposures, Porthcawl* [*SS 812 766*]; *Pant Mawr Quarry (type section of Pant Mawr Sandstone)* [*SS 828 802*]; *Stormy West Quarry (for Pant Mawr Sandstone)* [*SS 846 806*].

OYSTERMOUTH BEDS (OB)

The Oystermouth Beds (George and others, 1976), the highest formation of the local Dinantian succession (Figure 3), correspond to the Upper Limestone Shales of George (1933). They are present only in the west of the district, at Kenfig [SS 806 815] and to the north of Bridgend [SS 902 821]. Namurian overstep has removed them elsewhere.

The Oystermouth Beds comprise interbedded thin limestones and shales, silicified limestones and banded cherts which, in the Kenfig Borehole [SS 8055 8167] (Woodland and Evans, 1964; Appendix 2), yielded the goniatites *Neoglyphioceras* sp. and *Sudeticeras* sp., indicative of a late-Brigantian age (Ramsbottom 1954; George and others, 1976). The formation is poorly exposed within the Bridgend district and a complete sequence is not recorded, but it attains a thickness of up to 61 m to the west, in the Swansea district. Up to 21 m were encountered in the Kenfig Borehole, overlain, apparently conformably, by Namurian shales, but in faulted contact with the underlying Oxwich Head Limestone. Immediately north of the district, the uppermost 18 m of the formation are thrust over Namurian rocks in a motorway cutting at Sarn Park [SS 918 827]. The upper 7.5 m of the Oystermouth Beds at this locality, comprise silty mudstones with scattered thin argillaceous limestones, capped by 1.5 m of banded chert. The sharp contact with overlying Namurian shales appears to be in structural continuity, although the occurrence at this level of pebbly lags composed of chert intraclasts, phosphatic nodules, rolled bone fragments and coprolites is indicative of a stratigraphical non-sequence.

The Oystermouth Beds indicate a marked increase in the amount of terrigenoclastic mud supplied to the Dinantian shelf and a concomitant reduction in skeletal carbonate production; silicification also was evidently important. The formation reflects the demise of the carbonate regime that had predominated during the Dinantian and marks the transition into the siliciclastic deltaic environment of the Upper Carboniferous. There appears to have been an essentially conformable passage in the west of the district, but increasing nonsequence, culminating in unconformity, towards the east.

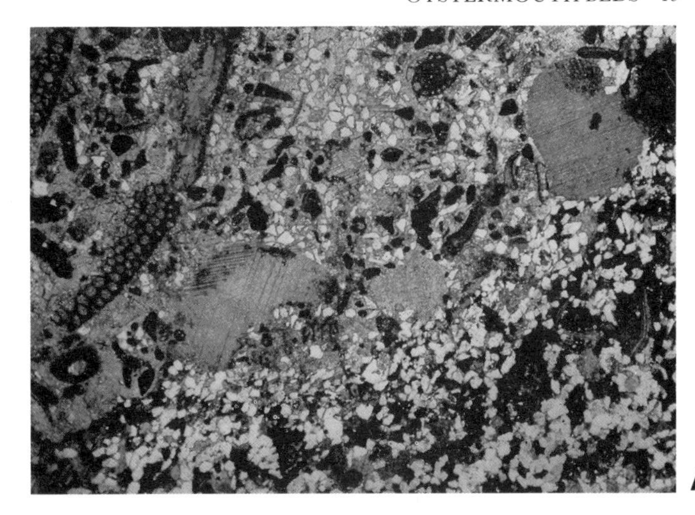

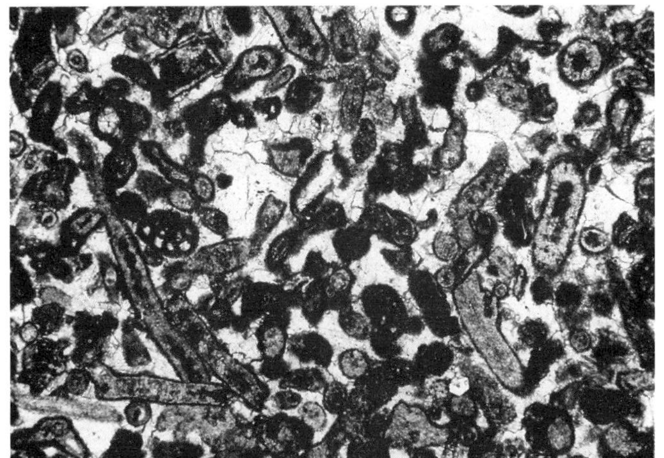

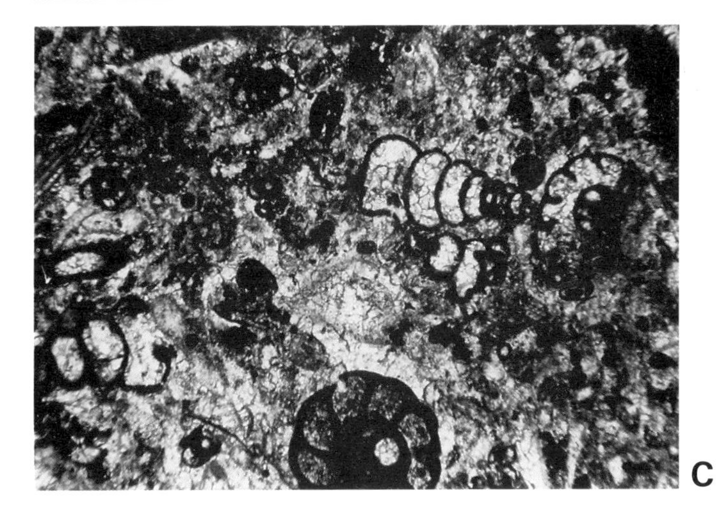

Plate 9 Dinantian limestone textures: Oxwich Head Limestone

A Fine-grained calcareous sandstone containing abundant skeletal debris; Pant Mawr Sandstone, Tythegston Quarry [SS 8523 7890]. Field of view 6.5 mm, cross polarisers (E 56490)

B Peloidal, skeletal packstone–grainstone; Oxwich Head Limestone, Gaen Quarry [SS 824 804]. Field of view 3.5 mm (ARE 2064)

C Foraminiferal packstone; Oxwich Head Limestone, Locks Common, Porthcawl [SS 8046 7754]. Field of view 4.5 mm (ARE 2034)

FOUR

Silesian (Upper Carboniferous)

INTRODUCTION

Upper Carboniferous (Silesian) rocks crop out in a narrow tract in the extreme north of the district, but are largely obscured by drift deposits. They range from Namurian to Westphalian in age and include the traditional subdivisions of Millstone Grit and Coal Measures. They comprise repetitive, cyclic sequences of siliciclastic sediments with, particularly in the Westphalian, coal seams, seatearths and ironstones, all being indicative of a variety of deltaic and coastal plain environments. Periodic transgressions are represented by dark mudstones with distinctive marine faunas (marine bands), which have been widely used in correlation.

The establishment of this facies complex in South Wales during the Silesian has been attributed mainly to uplift of a landmass (St George's Land) to the north, from which much of the sediment was derived (George, 1970; Owen, 1964, 1971; Kelling, 1974). The uplift of St George's Land and movement on major structures such as the Vale of Neath Disturbance and the Usk Anticline (George, 1956) have combined to produce sub-Namurian overstep and unconformity of variable magnitude.

MILLSTONE GRIT SERIES [NAMURIAN] (MG)

The Millstone Grit Series of the South Crop of the coalfield (Woodland and Evans, 1964) encompasses rocks of Namurian age and partly equates with the earlier 'Millstone Grit' division of Woodland and others (1957b). Namurian chronostratigraphy is largely based on the sequence of goniatite faunas found within marine bands (Ramsbottom and others, 1978); these show that the base of the Millstone Grit Series within the district is markedly diachronous, the basal beds ranging from Arnsbergian (E_2) in the west, to Marsdenian (R_2) in the east. The top of the division is conventionally taken at the base of the Subcrenatum Marine Band.

The lowest Namurian strata in the west of the district appear conformable with the underlying Dinantian. They were encountered in the Kenfig Borehole [SS 8055 8167] (Appendix 2), where they comprise up to 35 m of dark grey shales with numerous thin black cherts and sporadic coarse-grained sandstones, up to 1.5 m thick, overlying the limestones, shales and banded cherts of the Oystermouth Beds. Farther to the east, in the Sarn Park motorway cutting [SS 918 827], immediately north of the district (Figure 6), over 100 m of Namurian rocks are exposed. They dominantly comprise dark brown to black laminated siltstones and shaly mudstones, punctuated at intervals by thick sequences (locally up to 15 m) of medium- to coarse-grained, commonly pebbly, cross-bedded and planar-laminated sandstones. Nonsequence at the base of the Namurian in this area is indicated, in the lowermost beds, by the presence of thin pebbly lags of chert intraclasts, phosphatic nodules, abraded bone fragments and coprolites; this is confirmed by the lowest recorded faunas, which include the bivalve *Posidonia corrugata*, the conodont *Gnathodus bilineatus* and the goniatite *Cravenoceratoides nitidus*, indicating a mid to late Arnsbergian ($E_{2b2}-E_{2b3}$) age. The highest exposed strata in this section belong to the late Marsdenian (R_{2c}) stage, as indicated by the presence of the *Bilinguites superbilinguis* and *Donetzoceras? sigma* Marine Bands; the intervening Chokierian, Alportian and Kinderscoutian stages are all present to some extent, although not every faunal horizon is represented. The sequence at Sarn Park is similar to that formerly exposed at Litchard [SS 908 820], in the extreme north of the district (Woodland and Evans, 1964).

Evidence of the increasing eastward Namurian overstep has been recorded from the extensive underground workings at the Llanharry iron-ore mine (Williams, 1958), where black shales of the Marsdenian (R_2) stage fill irregularities and hollows in the underlying Oxwich Head Limestone. To the north-east of Groesfaen, at Creigiau Quarry [ST 088 821], strata of late Marsdenian (R_{2c}) age overlie dolomitised limestones of the Hunts Bay Group (Waters and Lawrence, 1987; Figure 6).

Locality: Stream section, Nant Heol-y-Geifr [SS 9537 8231].

COAL MEASURES [WESTPHALIAN]

Coal Measures crop out in the extreme north of the district, resting conformably on the Millstone Grit Series. They are subdivided lithostratigraphically into Lower, Middle and Upper Coal (or Pennant) Measures, all of Westphalian age. There are few exposures, but the sequence is known in detail from extensive drilling, opencast and deep mining operations carried out by British Coal.

The Coal Measures generally comprise mudstones, siltstones and sandstones, with associated economic coals and seatearths, typically arranged in sedimentary cyclothems (Woodland and Evans, 1964). Sandstones are most common in the lower parts of the sequence and locally, near the top, in the Llynfi Beds (Figure 7).

The Westphalian Series is divided into four stages (Westphalian A to D), on the basis of widely correlatable marine bands (Figure 7); only the lower three of these stages are represented in the district. The standard names (Ramsbottom and others, 1978) for the principal marine bands are used here although the local South Wales name, where it differs substantially, is given in brackets at first mention below.

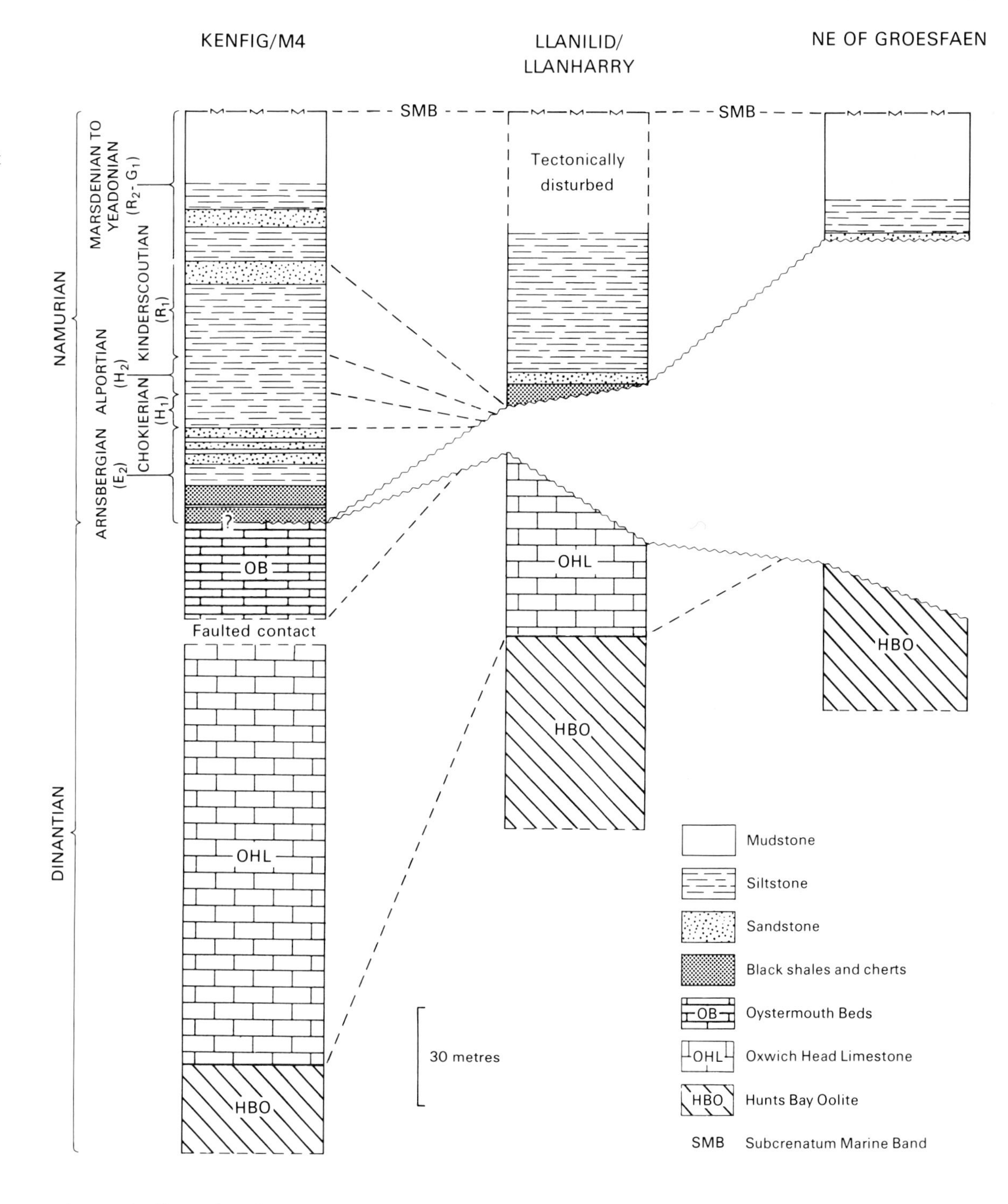

Figure 6 Comparative sections of the Namurian sequence of the Bridgend district

Lower Coal Measures [Westphalian A] (LCM)

The Lower Coal Measures of the South Wales Coalfield are the equivalent of the lowest stage of the Westphalian, encompassing strata ranging from the base of the Subcrenatum Marine Band to the base of the Vanderbeckei (Amman) Marine Band (Figure 7). The beds vary in thickness from about 360 m between Bridgend and Llanilid, to about 125 m in the east of the district; they are poorly exposed, but have been described from the adjacent Pontypridd district (Woodland and Evans, 1964). Massive and flaggy sandstones are commonly interbedded with the mudstones in this part of the sequence. The most prominent and laterally persistent is the Cefn Cribbwr Sandstone (Figure 7), up to 36 m thick, occurring above the Listeri Marine Band. The lowest persistent coal within the local Westphalian sequence is the **Garw** seam, up to 0.7 m thick. Above is a sequence of mainly argillaceous sediments, containing extensive and previously worked horizons of nodular and thinly bedded sideritic ironstones. Coal seams within these sediments display rapid thickness variations and splitting. The **Five-Feet** and **Gellideg** locally merge to give a 2 m thick seam; similarly, the **Seven-Feet** and **Yard** coals combine in places to form a single seam over 1 m thick. The siltstones and thin sandstones which

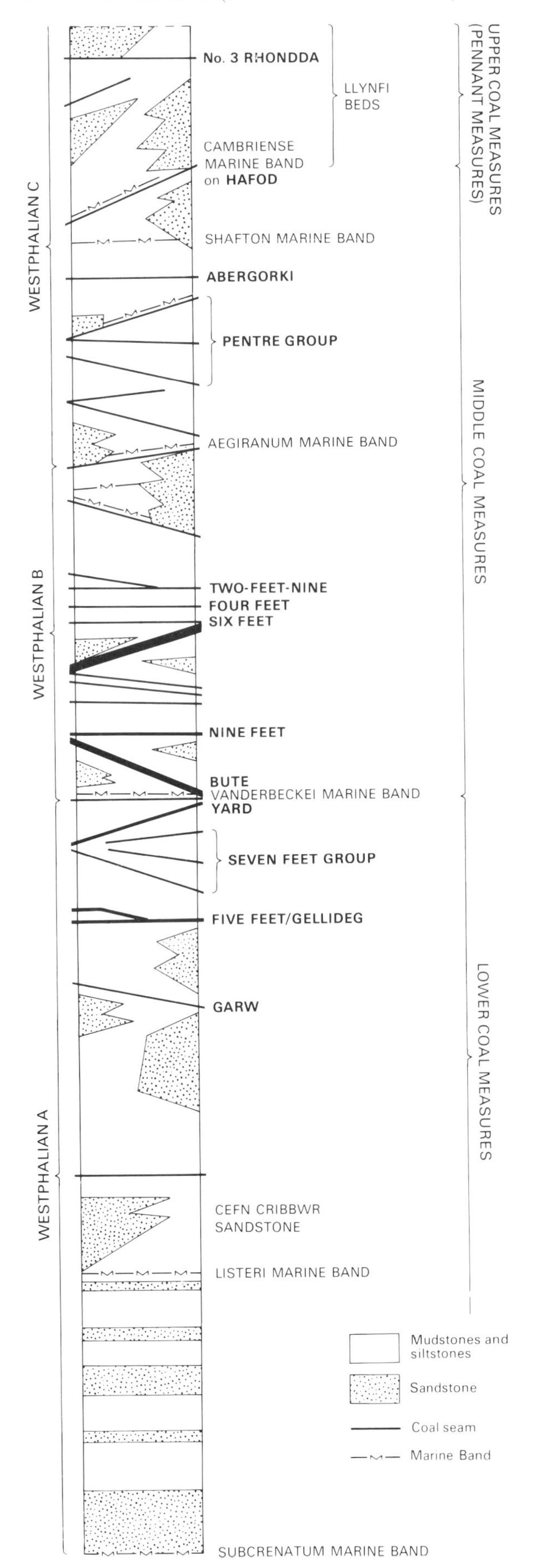

Figure 7 Generalised section of the Westphalian sequence of the Bridgend district (not to scale)

overlie the **Yard** coal are capped by a thin pyritic coal, the **Amman Rider**, immediately overlain by the Vanderbeckei Marine Band.

Localities: Quarries at Penprysg and Ty-draw [SS 9622 8250 and SS 9672 8229]; Llanilid Opencast Site [SS 993 819].

Middle Coal Measures [Westphalian B–C] (MCM)

The Middle Coal Measures of the South Wales Coalfield include strata between the base of the Vanderbeckei Marine Band and the Cambriense (Upper Cwmgorse) Marine Band. They range in age from Westphalian B to Westphalian C, the stage boundary occurring at the base of the Aegiranum (Cefn Coed) Marine Band, within the middle of the division. The Westphalian B stage of the Middle Coal Measures comprises a succession of coal-bearing cyclothems varying between 110 m and 140 m in thickness, which contain the thickest, as well as the most extensively worked coals in the district. Coal seams commonly divide and change thickness rapidly within this sequence. The **Bute** coal is about 2 m thick, with the thinner **Nine-Feet**, **Red Vein** and **Caerau** seams at intervals above. Important higher coals are the **Six-Feet**, **Four-Feet** and **Two-Feet-Nine**, all over 1 m thick. Within this mainly argillaceous sequence, fine-grained, massive, cross-bedded, quartzose sandstones, with channelled bases occur locally above some coals and in places cut down into the underlying seatearths. The upper part of Westphalian B contains the Haughton (Hafod Heulog) and the Sutton (Britannic) Marine Bands.

The Middle Coal Measures of Westphalian C age, overlying the Aegiranum Marine Band, comprise 85 m to 95 m of muddy siltstones and sandstones. Prominent coals within this sequence include the **Gorllwyn** seams which, in the Bridgend district, are relatively thin, the **Pentre** group of coals (1 m to 2.5 m thick) and the **Abergorki** seams. Above the latter is the Shafton (Lower Cwmgorse) Marine Band. The succeeding **Hafod** seam, over 1 m thick, is overlain by the Cambriense Marine Band.

Locality: Llanilid Opencast Site [SS 993 819].

Upper Coal Measures [Westphalian C] (UCM)

The Cambriense Marine Band marks the base of the sandstone-dominated Upper Coal Measures, traditionally known as the Pennant Measures, a sequence impoverished in marine bands but subdivided using widely recognised coal seams (Woodland and others, 1957a). Over 100 m of the lowest division, the Llynfi Beds, is present within the district, comprising dark siltstones with coarse-grained sandstones and a prominent coal, the **No. 3 Rhondda**.

Locality: Stream section, Nant Graean [ST 0140 8200 to ST 0161 8208].

FIVE

Triassic

INTRODUCTION

The culmination of the Hercynian Orogeny in South Wales, at the end of the Carboniferous, resulted in uplift, which was followed by extensive denudation throughout the Permian and into the Triassic. Permian deposits are absent from the district and the earliest preserved Triassic sediments are those of the Mercia Mudstone Group, comprising a 'marginal facies' of conglomerates, breccias and sandstones, passing laterally into, and locally interbedded with, a sequence of lacustrine siltstones and mudstones. Late in the Triassic, a marine transgression introduced the Penarth Group sediments into the district; in the east this sequence dominantly comprises marine mudstones with subordinate sandstones and limestones, which pass north-westwards into a marginal facies of sandstones and red mudstones indicative of a shallow marine and coastal plain setting (Figure 8).

The distribution of the sediments, particularly those of the Mercia Mudstone Group, was largely influenced by the contemporary topography, with the Dinantian limestone ridges of the eroded and breached Cardiff–Cowbridge Anticline providing a source for the breccias and conglomerates of the marginal facies. By the late Triassic, however, these features were so subdued that, although they affected the distribution of the Penarth Group facies (Francis, 1959), they contributed little clastic sediment to the succession.

The Triassic nomenclature and chronostratigraphical divisions used here follow those of Warrington and others (1980). The Mercia Mudstone Group of the district probably falls within the Norian stage, but palaeontological evidence is lacking. The overlying Penarth Group is Rhaetian; the position of the Norian–Rhaetian stage boundary is uncertain, but probably lies within the upper part of the Mercia Mudstone Group. The Rhaetian is the highest stage of the Triassic; its top is taken at the first appearance of the ammonite *Psiloceras*, which marks the base of the succeeding Hettangian stage of the Lower Jurassic. In the Vale of Glamorgan, the Rhaetian embraces the lowermost few metres of the Lower Lias, which is described in Chapter Six.

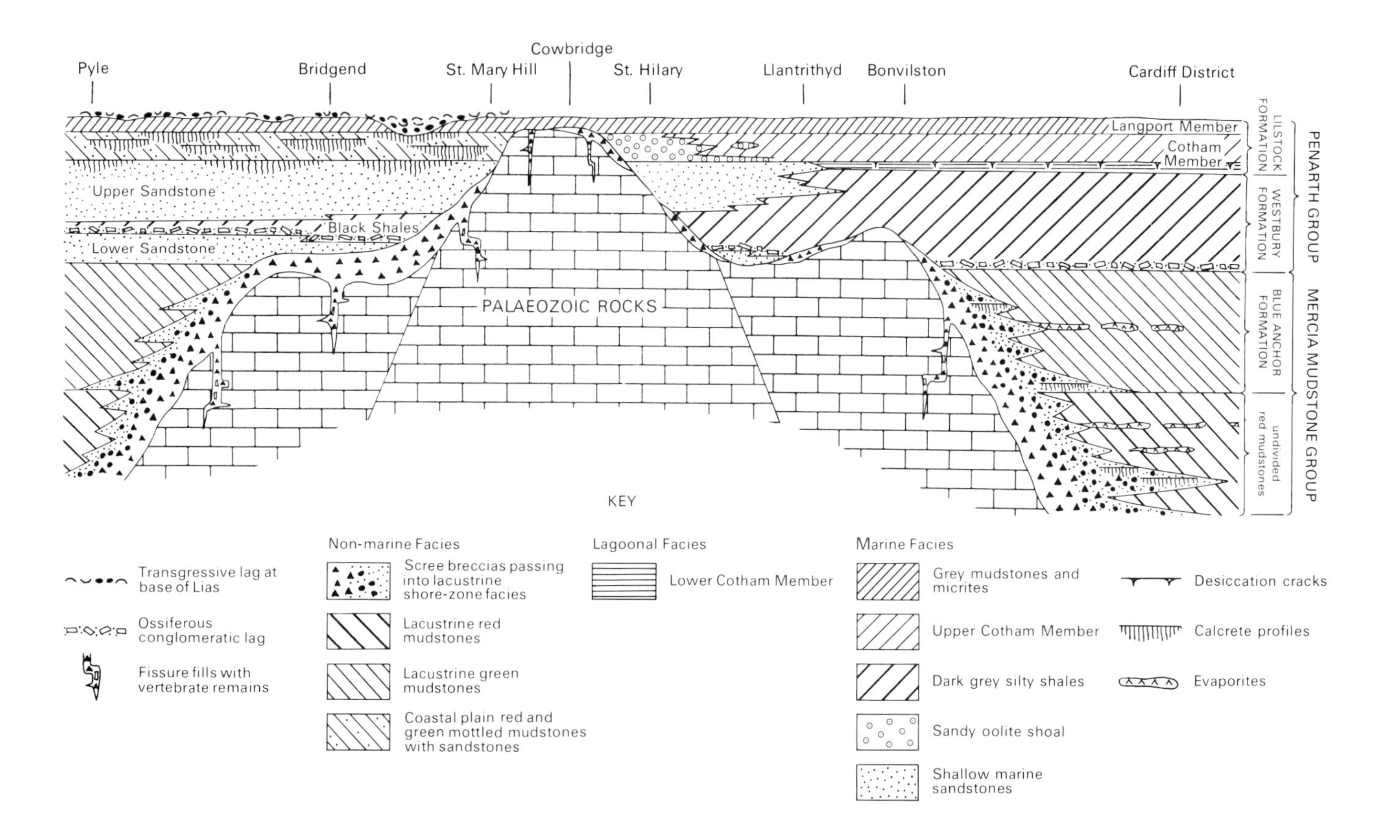

Figure 8 Schematic section through the Triassic succession of the Bridgend district (not to scale)

Plate 10 Triassic scree deposits resting with stepped unconformity on Dinantian High Tor Limestone; coast section at Ogmore-by-Sea [SS 8664 7400] (A 14288)

MERCIA MUDSTONE GROUP

The major part of the Mercia Mudstone Group in the Bridgend district is of a marginal facies, which fringes the Dinantian limestone ridges of the Cardiff–Cowbridge Anticline in the east and crops out extensively in the Coychurch–Coity area [SS 935 810], in the vicinity of Laleston [SS 860 795], and between Porthcawl [SS 830 775] and Kenfig Burrows [SS 795 810]. It generally rests with marked unconformity on the older rocks, a feature well displayed at Ogmore-by-Sea [SS 862 750] (Plate 10). Included with the marginal facies are Triassic sediments found within the numerous fissure and cavern systems that are well developed in the Dinantian limestones (Plates 11 and 12). The marginal facies passes laterally into a sequence of mudstones, comprising a lower division of red mudstones, overlain by the mainly green mudstones of the Blue Anchor Formation (Plate 13). The mudstones crop out intermittently between Kenfig Burrows [SS 790 820] and Hendre [SS 940 819] in the north of the district, and in the east, mainly in the vicinity of Clawdd-côch [ST 055 776], Pendoylan [ST 065 765] and Groesfaen [ST 075 803].

The Mercia Mudstone Group in the district is up to 85 m thick, including up to 14 m of the Blue Anchor Formation, but thicknesses are very variable. Despite its extensive outcrop, the marginal facies is rarely more than 25 m thick.

Marginal facies (M/mf)

This facies generally comprises a sequence of ill-sorted to fairly well-sorted breccias and conglomerates with clasts dominantly of Dinantian limestone in a reddened matrix of limestone fragments (Plate 14a); in places, the clasts are cemented by thinly laminated, buff micrite. The clasts vary from angular to rounded; large slabs of Dinantian limestone occur locally within the poorly sorted breccias. Bedding is only poorly developed within the conglomerates and breccias

Plate 11 Swallow-hole in Dinantian Black Rock Limestone filled with Triassic sediment. Longlands Quarry [SS 9274 7720] (A 14283)

Plate 12 Slickensided calcite speleothem (cave deposit) within a Triassic fissure; Ewenny Quarry [SS 0926 7675] (A 14280)

Plate 13 Transition from the undivided red mudstones into the Blue Anchor Formation, the contact being taken at the highest red mudstone bed; Mercia Mudstone Group, disused railway cutting near Pyle [SS 8255 8167]

Beds of concretionary limestone are present within the greyish green mudstones of the Blue Anchor Formation and at a lower level in the red mudstones (A 14357)

(Plate 10); bedding surfaces are generally irregular and locally erosional, and cross-bedding has been recorded in a few places. Subordinate thin beds of calcarenite, sandstone, siltstone and red mudstone, locally containing calcrete nodules, occur at intervals throughout the sequence. The marginal facies of the Blue Anchor Formation, although not distinguished separately on the map, comprises a distinctive sequence of pale greenish grey, fine-grained, limestone-chip breccias and conglomerates, interbedded with pale green and buff, porcellaneous calcite mudstones, subordinate nodular dolomitic horizons and fenestral carbonates (Plate 14b).

Triassic deposits fill an extensive network of fissures and caverns in the Dinantian limestones. These vary in size from open joint fissures, a few centimetres wide, to deep potholes many metres in diameter and extending to depths of over 30 m (Plate 11). The deposits are generally of red and green mudstones and siltstones or, less commonly, of sandstones and conglomerates. Discoveries of Norian vertebrate remains in these sediments are summarised by Ivimey-Cook (1974).

The marginal facies represents a range of continental depositional environments. The poorly sorted breccias have

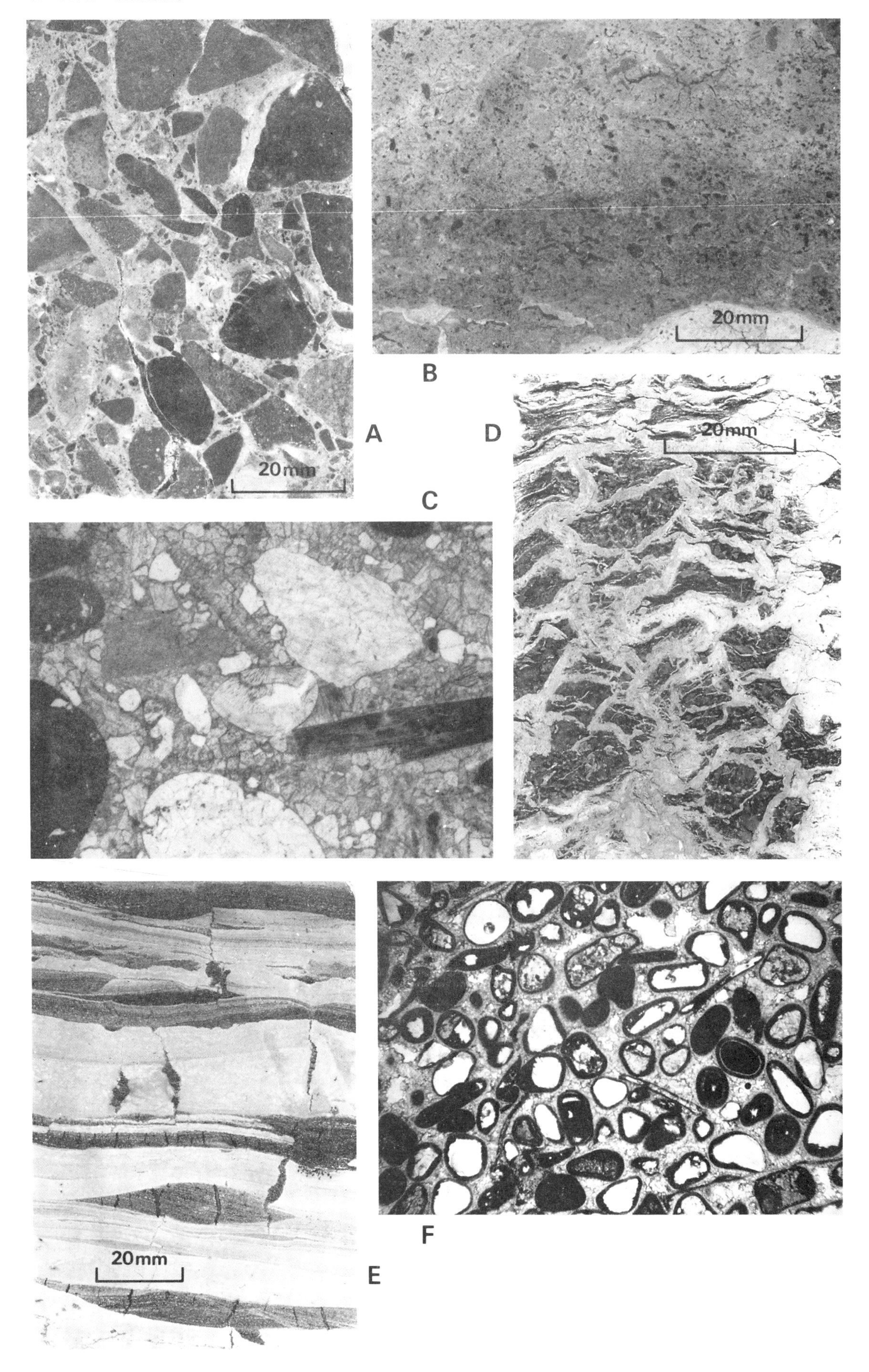
20mm
B
A
D
20mm
20mm
C
F
20mm
E

Plate 14 Mercia Mudstone Group and Penarth Group: textural and lithological variations

A Breccia composed dominantly of clasts of Dinantian limestones, in a matrix of progressively finer limestone fragments, cemented by buff micrite; Mercia Mudstone Group, marginal facies, near Tythegston [SS 858 793]

B Fine-grained limestone-chip breccia with a calcite mudstone matrix in which spar-filled desiccation fenestrae are developed; Blue Anchor Formation, marginal facies, near Howe Mill Farm [ST 0018 7244]

C Calcareous sandstone with shell and bone fragments, and pebbles of quartzite, phosphate and limestone; Westbury Formation (Upper Sandstone), near St Mary Church Station [ST 0198 7129]. Field of view 4.5 mm (E 58131)

D Green mudstones with a latticework of fine-grained buff dolomite (caliche) resulting from pedogenic processes; top of Westbury Formation (Upper Sandstone), Bridgend Hospital borehole [SS 911 811]

E Laminated calcite mudstone with lenticular, ripple cross-laminated fine-grained sandstones; Cotham Member, Gigman Mill [ST 0115 7181]. Vertical desiccation cracks within the mudstones are filled with sandstone. The cracks within the sandstones are spar-filled and result from tensional effects produced during differential compaction of the sands and muds

F Superficial ooid grainstone, with cores of quartz sand, skeletal grains and peloids; Cotham Member, near St Mary Church Station [ST 0210 7114]. Field of view 4.5 mm (E 58126)

been interpreted as subaerial scree deposits, whereas the well-sorted conglomerates are thought to represent the alluvial fans of ephemeral streams formed during periodic rainstorms in a semiarid climate (Bluck, 1965; Tucker 1977). The range of lithofacies probably reflects the degree of reworking by these streams as they flushed out the accumulated deposits from valleys incised into the Dinantian limestones. The interbedded sandstones and calcarenites probably developed on, and in front of, the alluvial fans as a result of sheet-floods during the rainstorms (Tucker, 1977; Waters and Lawrence, 1987).

The passage from the continental marginal facies into the lacustrine mudstone facies of the Mercia Mudstone Group is recorded in the transitional deposits of a narrow lacustrine shore-zone (Tucker 1978), which is characterised by interdigitation of the two lithofacies, with evidence of a fluctuating lake-level being recorded by calcrete nodules, dessication fenestrae, erosional benches and replaced evaporites.

Localities: Coastal sections at Sker Point [SS 787 797] and Ogmore-by-Sea [SS 862 749]; disused quarry, Pant-y-gog [SS 9592 7982]; road cutting at Ruthin [SS 973 800]; disused quarry, S of Clawdd-côch [ST 0565 7707]; Llantrisant Road cutting [ST 0515 8208]. Triassic fissure deposits occur at Pant-y-ffynnon Quarry [ST 045 740], Longlands Quarry [SS 927 772] (Plate 11), Ewenny Quarry [SS 093 675] (Plate 12) and Garwa Farm Quarry [SS 980 787].

Undivided red mudstones (MMG)

This facies comprises a monotonous sequence of reddish brown, massive, unfossiliferous, calcareous or dolomitic mudstones (Plate 13); thin bands of greenish grey mudstone occur at intervals and green spots and mottling are sporadically developed. Subordinate thin beds of red and green calcareous siltstone are present locally. Gypsum occurs as nodules, scattered throughout the sequence, and in nodule horizons.

The red mudstones represent the deposits of what was once a large, shallow, intermittent playa lake in which subaqueous deposition was punctuated by regressive periods, during which the lake level fell and there was extensive subaerial exposure. The calcareous siltstones within the sequence may represent shore-zone sediments, transported and deposited during these shallowing episodes, whilst intensive evaporation led to the formation of the gypsiferous horizons.

Locality: Disused railway cutting, North Cornelly [SS 8256 8172].

Blue Anchor Formation (BAn)

The Blue Anchor Formation (the 'Tea Green Marls' of previous authors) overlies the red mudstones with a fairly sharp contact (Plate 13) but, in the vicinity of Bonvilston [ST 062 728], Pendoylan [ST 061 767] and Groesfaen [ST 075 806], it oversteps them to rest directly on the marginal facies. The formation typically comprises up to 14 m of green and greyish green mudstones, often thinly laminated, frequently dolomitic, and locally interbedded with thin dolomites and dolomitic limestones; bands of reddish brown mudstone occur at intervals and gypsum nodules are common at a few horizons. Several mudstone beds exhibit dessication cracks and autobrecciation effects. Fish, plant and reptilian remains have been recorded from the upper part of the formation in Glamorgan (Ivimey-Cook, 1974), and the uppermost beds yield sparse palynomorph assemblages of organic microplankton and miospores (Orbell, 1973; Warrington, *in* Waters and Lawrence, 1987), which have Rhaetian affinities.

The Blue Anchor Formation lithofacies appears to be largely lacustrine, with dessication cracks and evaporites indicating periods of low lake level. However, the plant debris and spores in the upper part of the formation indicate a major climatic change at this time from arid to relatively humid conditions. This climatic amelioration may also be reflected in the widespread colour change from mostly red to mostly green sediments. The climatic changes are probably the result of marine influences which culminated in the late-Triassic marine transgression.

Locality: Disused railway cutting, North Cornelly [SS 8256 8172].

PENARTH GROUP (PnG)

The Penarth Group (the 'Rhaetic Beds' of previous authors) comprises a sequence of marine shales with subordinate sandstones and limestones in the east of the district, which pass north-westwards into a marginal facies, dominantly of sandstones and red mudstones. It disconformably overlies the Blue Anchor Formation, locally overlapping onto marginal facies of the Mercia Mudstone Group and, in places, resting unconformably on Dinantian limestones.

In the east, the Penarth Group, comprising the Westbury Formation overlain by the Lilstock Formation, is similar to that exposed at its eponymous type section in the Cardiff district (Warrington and others 1980; Figure 8) but, with a combined thickness of only 8 m, the two formations are not differentiated on the map. The Westbury Formation comprises dark grey, bioturbated, silty, locally pyritous shales with scattered thin layers rich in bivalves, including the characteristic bivalve *Rhaetavicula contorta*, and subordinate thin beds of calcareous sandstone and argillaceous and coquinoid limestone. In places, the base of the formation is marked by a distinctive 'bone-bed', usually a thin conglomeratic sandstone, containing abundant vertebrate bones and fish scales; the top of the formation is a sharp, locally erosive surface.

The Lilstock Formation is divided into the Cotham Member, overlain by the Langport Member. The former comprises greyish green calcareous mudstones and siltstones with, in the upper part, beds and lenses of sandstone containing peloids, ooids and shell debris which, in places, grade into sandy limestones (Plate 14f); these latter lithologies become dominant in the the St Hilary area [SS 015 723], locally forming beds up to 0.5 m thick. Planar- and cross-lamination and wave-ripple marks are common in the siltstones and sandstones (Plate 14e). The mudstones exhibit fenestral fabrics and polygonal dessication cracks up to 0.1 m deep, filled with material from the overlying bed; a widely recognised dessication horizon occurs in the middle part of the member, overlying a mudstone containing synsedimentary faults and slumps (Mayall, 1983; Waters and Lawrence, 1987). The Langport Member comprises a lower sequence of calcite mudstones and subordinate skeletal packstones with shaly partings (formerly the 'White Lias' or Langport Beds; Richardson, 1911), overlain by calcareous mudstones with thin, lenticular, planar- or cross-laminated sandstones and shelly, argillaceous limestones (formerly the Watchet Beds of Richardson, 1911, or the Watchet Member of Ivimey-Cook, 1974; but see Whittaker, 1978). Strata assigned to the Langport Member also occur in the area around Bridgend, where they form the only part of the Penarth Group not replaced by marginal facies (Figure 8). The top of the Penarth Group, at its type locality, is conventionally taken at the junction with the Paper Shales at the base of the Blue Lias; this horizon also marks the top of the Langport Member, as redefined by Warrington and others (1980). The contact appears sharp and conformable in the east of the district, but becomes disconformable to the north-west.

The fauna of the Penarth Group, although of limited diversity, is considerably more abundant and varied than that of the Blue Anchor Formation and is generally indicative of shallow marine environments, with periodic freshwater influxes. Marine conditions were initially established in the east of the district, the 'bone-bed' at the base of the Westbury Formation representing a transgressive lag or strand-line deposit (Sellwood and others 1970). The shales of the Westbury Formation were deposited in a relatively quiescent marine environment; periodic turbulent episodes, due to storm activity, are represented by the thin sandstones, cocquinoid limestones and shelly lags.

Shallowing, with local erosion at the top of the Westbury Formation, established restricted lagoonal conditions to the south and east. In this environment the lower part of the Cotham Member (Lilstock Formation) was deposited, with a fauna greatly reduced in numbers and diversity. The bed containing slumps and synsedimentary faults within the mudstones records tectonic activity (Mayall, 1983) contemporaneous with regional uplift and emergence, which led to the formation of the widespread horizon of dessication cracks. Marine conditions, with periodic freshwater influxes, were re-established in the upper part of the Cotham Member (Waters and Lawrence, 1987). The sandy, peloidal and oolitic limestones in this unit are interpreted as the remnants of a lagoonal barrier, driven shorewards in response to marine transgression; the thickest development of this facies, in the St Hilary area, is thought to record impingement of the barrier against an inherited topographical high, which resulted in its localised aggradation (Figure 8).

Continuing transgression finally introduced marine conditions into the west of the district, with the deposition of the Langport Member; the lithologies and fauna of this division are indicative of shallow, subtidal environments subject to periodic storms.

Localities: Stream section, Maendy Farm [ST 0773 7814]; disused railway cutting, south of St Mary Church Station [ST 0211 7112].

Penarth Group marginal facies (PnG/mf)

From Pyle [SS 823 823] to St Mary Hill [SS 969 785], the Penarth Group is largely represented by a marginal facies, dominantly comprising a thick sequence of sandstones (**S**) overlain by mottled green and red mudstones (**PnG/mf**); the sandstones persist south-eastwards into the St Hilary–Llantrithyd area (Figure 8). The marginal facies is unique in South Wales and its distribution was probably controlled by the pre-existing topography. It locally attains thicknesses of 20 m, with the lateral correlatives of the Westbury Formation being represented by the arenaceous sequence and the Lilstock Formation broadly equating with the overlying mudstones.

The marginal facies of the Westbury Formation comprises two sandstone units (the Lower and Upper Sandstones of Francis, 1959), with up to 2 m of intervening dark grey shales (the Black Shales of Francis, 1959; Figure 8). The Lower Sandstone rests with a sharp, erosional base on the Mercia Mudstone Group, and commonly fills hollows in the irregular Triassic topography. It comprises white, yellow and buff, medium- to coarse-grained, massive to cross-bedded and ripple-marked quartz sandstones, containing small quartz pebbles, wood fragments and fish remains, which are locally concentrated in thin, lenticular conglomeratic bands; bivalves have previously been recorded from a locality at Coity [SS 929 806] (Francis, 1959). On Stormy Down [SS 851 813] the Lower Sandstone is over 6 m thick (Francis 1959), but it thins progressively eastwards as it is overlapped by the Black Shales and Upper Sandstone; east of Bridgend it is only locally preserved, and it is largely absent south-east of Coychurch [SS 940 795].

The top of the Lower Sandstone is marked by an uneven,

erosional surface, which is sharply overlain by a thin, but locally persistent conglomerate containing pebbles of quartz, jasper and chert and abundant vertebrate remains. The succeeding Black Shales are identical to those of the typical Westbury Formation. They are dark grey and green, with thin, nodular limestones containing fish and vertebrate remains, and layers rich in bivalves, *R. contorta* being especially abundant. Laminae of siltstone and fine-grained sandstone are common in the upper parts, marking the transition into the Upper Sandstone.

The Upper Sandstone (known locally as the 'Quarella Stone', from the quarry of that name in Bridgend [SS 9038 8082]) is generally between 7 and 11 m thick in the north-west of the district, but thins markedly in the vicinity of Tythegston [SS 857 789]. To the south-east of Bridgend it thins progressively, being only 5 m thick near St Hilary and less than 1 m near Llantrithyd [ST 044 730]; south and east of Llantrithyd, it passes laterally into the shales of the Westbury Formation. It has a gradational lower contact with the Black Shales in the north-west of the district, but oversteps them east of Bridgend, to rest on older rocks. The transition into the typical Westbury Formation in the south-east of the district is complex, probably involving rapid interdigitation of the two facies (Figure 8). The Upper Sandstone generally comprises a basal division of brown and buff, fine- to medium-grained, thinly bedded quartz sandstones, interbedded with subordinate thin, black micaceous siltstones and shales; flaser and linsen bedding commonly occur within the sandstones. These are succeeded by pale grey, yellow and buff, medium-grained, planar-, low-angle and hummocky cross-laminated sandstones, with scours and flame structures, defined by carbonaceous streaks and micaceous partings. Thin conglomeratic lags, commonly of intraformational mudstone clasts, locally form the erosive bases of subordinate beds of massive sandstone, which fine upwards into thinly laminated sandstones and pale greenish grey, micaceous, silty mudstones. Bivalves, including *R. contorta*, occur throughout, but are particularily abundant in the basal part; bioturbation is common. The upper part generally comprises greenish yellow and buff, massive sandstones, subordinate thinly bedded sandstones with shale partings and thin layers of bivalve coquina; in the St Hilary area, these coquinas are interbedded with thin pebbly sandstones and breccias containing clasts of Dinantian limestones, chert, mudstone, quartzite and vertebrate remains (Plate 14c). Brecciated micritic and dolomitic nodules are recorded at the top of the Upper Sandstone in boreholes near Bridgend, forming a gradational passage into the overlying marginal Lilstock Formation (Figure 8; Plate 14d).

The marginal facies of the Lilstock Formation comprises up to 6 m of mottled bluish green, purplish red and buff calcareous mudstones, with subordinate thin lenses of siltstone and fine sandstone. Horizons of buff, nodular, dolomitic, micritic and sandy limestone, with thin pockets and lenses of variegated sandstone occur in places, veined and interbedded with green mudstones. These marginal facies are overlain by beds assigned to the Langport Member.

The arenaceous facies represents shoreface sediments deposited around low islands. The sandstones may have originally been derived from the Silesian rocks to the north (Francis, 1959; Crampton, 1960), but they have subsequently undergone marine reworking. The erosion surface on which the Lower Sandstone rests indicates that a period of regression occurred at the top of the Blue Anchor Formation. The sandstone was deposited in a shallow sea, subject to periodic storm activity. It was introduced during the transgression at the base of the Westbury Formation, and for this reason is only preserved in depressions and where it aggraded against shorelines.

The transgressive passage of a high-energy strandline is recorded by the ossiferous conglomeratic lag at the top of the Lower Sandstone, comparable to the 'bone-bed' in the east. The subsequent deposition of the Black Shales over much of the district represents the acme of this transgression. Much of the pre-existing topography was probably inundated by this time, although some areas remained emergent, notably in the vicinity of St Mary Hill, and the Dinantian massif between Ogmore-by-Sea and Cowbridge.

The Upper Sandstone is interpreted as a progradational sequence which developed during stillstand at the end of the transgression. It shows overall coarsening-upwards from interbedded offshore shales and shoreface sands through sandstones in which the increasing role of storms is indicated by the frequency of massive beds and their pebbly lags. The culmination of this progradational event is represented by the massive sandstones and interbedded cocquinas at the top of the Upper Sandstone. Progradation of the Upper Sandstone was generally from the north-west, although locally it may have aggraded around small islands and on shoals. One of these islands was probably the source of the thin breccias at the top of the sandstone in the St Hilary area.

The subsequent regression, which established lagoonal conditions in the south-east, resulted in emergence at the top of the Upper Sandstone in the north-west. Pedogenic alteration at this level is indicated by the dolomitic nodules, which are interpreted as calcrete profiles (Figure 8; Plate 14d). The succeeding red and green mudstones represent coastal plain overbank deposits in which pedogenic processes gave rise to irregular horizons of nodular dolomite. Their aggradation was probably a response to rising base-level during the late-Cotham transgression, prior to their eventual inundation late in the Triassic (Figure 8).

Localities: M4 motorway cutting at Stormy Down [*SS 845 810*]; *cutting at Brackla Industrial Estate* [*SS 9172 8145*]; *road cutting, south of Coity* [*SS 9290 8056*]; *disused quarry, east of Dyffryn* [*SS 9514 8158*]; *crags, west of St Mary Hill* [*SS 9521 7920*]; *R. Thaw valley* [*ST 0200 7120*].

SIX

Jurassic

INTRODUCTION

The late Triassic and early Jurassic marine transgression covered much of the district and resulted in the deposition of the Lower Lias, a thick sequence of interbedded limestones and mudstones (frontispiece). The Triassic uplands, mainly remnant Carboniferous massifs, were gradually inundated but a marginal (shoreline) facies of conglomerates, calcarenites and, locally, oolites (Figure 9) was sustained around the massifs throughout much of Lower Lias times, in part by contemporaneous uplift along the Vale of Glamorgan Axis. A temporary progradation of the marginal facies in the upper part of the sequence may reflect a change in the relative rates of uplift and drowning.

The majority of the Lower Lias of the district belongs to the Hettangian and Lower Sinemurian stages of the Lower Jurassic (Figure 9). It is currently accepted (Cope and others 1980) that the base of the Jurassic in the British Isles should be taken at the base of the *planorbis* Subzone, defined by the earliest appearance of the ammonite *Psiloceras*. This occurs about 5 m above the base of the Lower Lias in the Vale of Glamorgan (Waters and Lawrence, 1987); the lowermost beds are thus of Triassic age but, for convenience, they are dealt with here.

The original extent the Lower Lias is unknown, but it has been suggested that these rocks formerly covered the coalfield to the north (Wobber, 1966; Owen, 1967). They underlie much of the Bristol Channel, being overlain by Middle and Upper Jurassic rocks preserved in broad synclinal structures.

LOWER LIAS (LLi)

The Lower Lias comprises up to 150 m of thinly interbedded limestones and calcareous mudstones, a facies known as the Blue Lias (Cope and others, 1980) and widely recognised throughout southern Britain. The limestones are generally bluish grey, planar-, wavy- and nodular-bedded, argillaceous wackestones, containing variable amounts of skeletal debris, locally as winnowed lags. They have a diverse, though scattered macrofauna of uncompacted brachiopods, bivalves, gastropods, ammonites and echinoderms. Bioturbation is a common feature and includes the large branching burrow systems of *Thalassinoides* occurring on bedding surfaces. The mudstones are usually dark grey, shaly and bioturbated; invertebrate remains are similar to those in the limestones, but are commonly crushed.

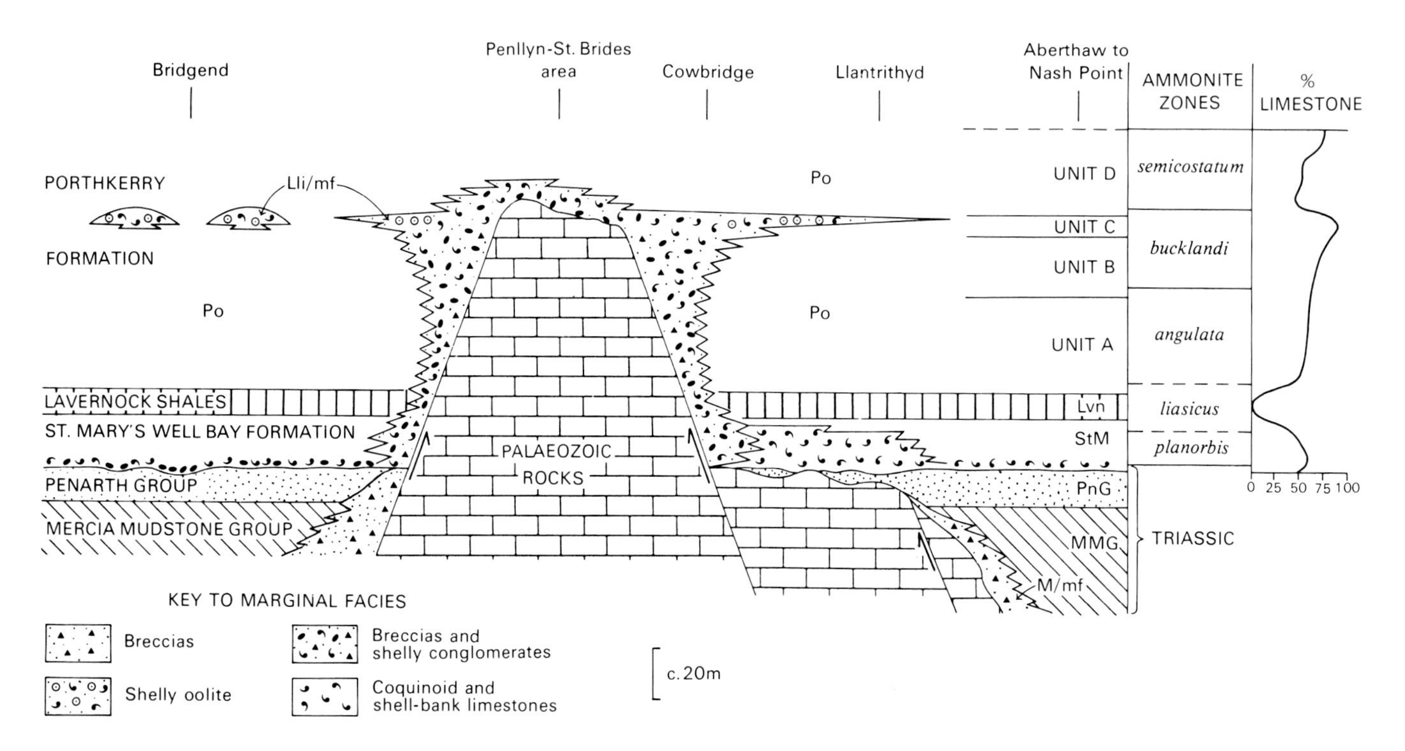

Figure 9 Schematic section through the Lower Lias of the Bridgend district, illustrating the relationships between the Blue Lias and its marginal facies. Variations in the percentage of limestone within the Blue Lias are indicated (those for Unit D are based on the section at Dunraven)

The Blue Lias was deposited in a shallow epeiric sea. The rhythmic alternations of limestone and shale may be primary (Hallam 1960), but it is generally accepted that considerable diagenetic modification has taken place (Hallam 1964, Sellwood 1970). The effects of pressure solution, in particular, are ubiquitous; many of the mudstone partings and much of the nodularity displayed by the limestones are the result of this process. The limestones may represent diagenetically altered accumulations of clastic sediment (Anderton and others 1979), with winnowed shell concentrates having been deposited within the limits of storm wave-base, and much of the finer-grained carbonate produced by coccolithoporids living within marine surface waters. An alternative view is that the limestones were precipitated from carbonate-enriched waters (Hallam, 1960; 1964). A study of similar, but less modified limestone and shale alternations (Sellwood, 1970) indicated that each rhythm represented a rapid deepening, followed by a slow regressive event during which mudstone deposition gave way to condensed limestones with winnowed skeletal lags deposited within storm wave-base; it has been suggested that these rhythms may reflect the interplay of eustatic sea-level fluctuations and local tectonic subsidence.

The Blue Lias of the Vale of Glamorgan has been subdivided into three formations (Waters and Lawrence, 1987), based on ratios of limestone to mudstone; these are, in ascending order, the St Mary's Well Bay Formation, the Lavernock Shales and the Porthkerry Formation (Figure 9). The St Mary's Well Bay Formation encompasses strata ranging from the base of the Pre-planorbis Beds into the *Alsatites liasicus* Zone of the Hettangian. The overlying Lavernock Shales and lowermost Porthkerry Formation are probably of *A. liasicus* Zone age. The lower part of the Porthkerry Formation largely falls within the *Schlotheimia angulata* Zone, but the bulk of the formation on the coast is assigned to the overlying *Arietites bucklandi* and *Arnioceras semicostatum* Zones of the Lower Sinemurian. The highest beds of the Porthkerry Formation, containing *Euagassiceras resupinatum* Subzone faunas from the upper part of the *A. semicostatum* Zone, have been recorded near Bridgend (Cope and others, 1980; Hodges, 1986).

St Mary's Well Bay Formation (StM)

This formation commonly forms a low escarpment above the Penarth Group, but is generally poorly exposed. In the north of the district it crops out intermittently between St Mary Hill [SS 967 783] and Pyle [SS 826 823]. In the east, it occurs near Groesfaen [ST 070 810] and in the vicinity of Pendoylan [ST 060 767]. In the south-east, it crops out extensively between Llanmihangel [SS 982 719], Penmark [ST 059 689] and Bonvilston [ST 065 740].

The formation comprises up to 20 m of limestones and mudstones in approximately equal proportions. It rests conformably on the Penarth Group in the south and east of the district, where its base is defined as the base of the Paper Shales (Richardson, 1905) which comprises 0.2 m of fissile siltstones and silty mudstones. In the north-west the base of the formation appears disconformable, in places being marked by a thin conglomeratic lag containing clasts derived from the underlying Penarth Group marginal facies (Figure 9; Plate 17a). The top of the formation is gradational, although in the adjacent Cardiff district it is taken at the top of a prominent limestone bed (Bed 86 of Waters and Lawrence, 1987).

The lower 3 to 4 m of the formation comprises a widely recognised sequence of planar-bedded limestones with thin mudstone partings, containing abundant bivalves, notably *Liostrea hisingeri* and *Modiolus*, commonly in winnowed coquinas and bioclastic lags (the '*Ostrea* Beds' of previous authors); these beds, together with the underlying Paper Shales, comprise the Bull Cliff Member (Waters and Lawrence, 1987). The Planorbis Mudstones and Lower and Upper Laminated Beds are distinctive units that occur above the Bull Cliff Member (Trueman, 1920; Hallam, 1960; 1964). The Planorbis Mudstones (Bed 30 of Trueman 1920), with numerous casts of *Psiloceras planorbis* on bedding surfaces, occur 6 to 7 m above the base of the formation. The Lower Laminated Beds, 8 to 9 m above the base, consist of one or more planar-bedded, laminated limestones, 0.4 m thick, which in places pass laterally into mudstones. The Upper Laminated Beds, 0.9 m thick, comprise two to three planar-bedded, laminated limestones with mudstones, which occur 10 to 11 m above the base of the formation. The upper parts of the formation near St Mary Church [ST 001 716] have yielded *Caloceras johnstoni*, of the *johnstoni* Subzone; the presence of *Waehneroceras* in the Cardiff district (Waters and Lawrence, 1987), indicates that the higher beds are of the *liasicus* Zone.

Localities: Disused quarry, south-west of Tre-Aubrey [ST 0295 7198]; old stone pit, north-west of Brackla Hill [SS 9136 3081].

Lavernock Shales (Lvn)

The Lavernock Shales (Strahan and Cantrill, 1904; Trueman, 1920; Waters and Lawrence, 1987) are a sequence of dark grey, calcareous, bioturbated, shaly mudstones, with subordinate thin beds of nodular limestone and variable amounts of skeletal material. They are 12 m thick in the south-east and 9 to 10 m around Bridgend, but thin markedly near Whitmore Stairs [SS 898 713] (Plate 15) and Llandough [SS 987 729], where they are replaced by a marginal facies. They are generally poorly exposed; over most of the district they crop out on steep, wooded slopes above the St Mary's Well Bay Formation, giving rise to poorly drained, boggy ground with small landslips, numerous springs and associated deposits of calcareous tufa. Exposures of the top of the Lavernock Shales between Temple Bay [SS 890 725] and Whitmore Stairs have yielded *Laqueoceras* cf. *sublaqueus*, *Lytoceras articulatum*, *Waehneroceras* and *Caloceras*, indicative of the upper (*laqueus*) Subzone of the *liasicus* Zone.

Localities: Old Ordnance Factory cuttings, Brackla Hill [SS 9146 8052 to SS 9215 8084].

Porthkerry Formation (Po)

The Porthkerry Formation crops out extensively in the area around Bridgend and over the broad tableland in the south of the district, with smaller outcrops in the dissected core of

Plate 15 The uppermost part of the Lavernock Shales, underlain by Lower Lias marginal facies calcarenites on the foreshore and in the shallow dome at the base of the cliffs. Whitmore Stairs [SS 8980 7127] (A 14350)

The Porthkerry Formation (Units A to C) overlies the Lavernock Shales in the cliff, with the highest beds formed of the amalgamated limestones of Unit C

Plate 16 The upper part of Unit B and Unit C of the Porthkerry Formation; Nash Point [SS 9148 6840]. Height of cliffs approximately 32 m (A 14368)

C—limestones of Unit C; Si—Silicified Bed; M—Main Limestone; SB—Sandwich Bed

the Cardiff–Cowbridge Anticline; the spectacular cliffs of the Glamorgan coast (frontispiece) are largely of this formation. It comprises up to 120 m of interbedded fine-grained skeletal wackestones and calcareous shaly mudstones similar to the St Mary's Well Bay Formation, and containing a varied fauna of bivalves, including *Pinna*, and *Plagiostoma*, with echinoderms, ammonites and gastropods; 'nests' of *Gryphaea* (Plate 17b) are common in the shales and the large branching burrow-systems of *Thalassinoides* are widespread.

The formation overlies the Lavernock Shales with a gradational contact; no higher strata are preserved in the Vale of Glamorgan. The presence of conspicuous and laterally extensive marker beds have allowed detailed correlation of coastal exposures. Four informal lithostratigraphical units, which differ markedly in their gross bedding characteristics and limestone–mudstone ratios, have been recognised in the cliff sections; these are, in ascending order, Units A to D (Figure 9).

Unit A has a gradational contact over about 2 m, with the underlying Lavernock Shales. It comprises about 36 m of thin, impersistent, nodular- and wavy-bedded limestones and interbedded mudstones (frontispiece and Plate 15) in ratios of about 55:45 (Figure 9). The top of the unit is taken at a conspicuous mudstone (Bed 32 of Waters and Lawrence, 1987). The unit probably ranges from the upper part of the *liasicus* Zone to the upper *angulata* Zone, the latter being confirmed by the occurrence of *Schlotheimia* sp. at Temple Bay [SS 890 725].

Unit B comprises 26.5 m of massive and planar-bedded limestones, commonly in amalgamated beds 1 m or more thick, with thin mudstone partings (Plate 16). Limestone–mudstone ratios increase upwards, through the unit, from about 60:40 near the base, to 80:20 in the upper parts (Figure 9 and frontispiece). Silicification is locally important, particularly near the top, where many fossils are beekitised and chert nodules are common. A widely recognised 0.7 m limestone (Bed 55 of Trueman 1930) 1 m below the top of Unit B has been informally termed the 'Silicified Bed' (Plate 16); its lower half exhibits pervasive and laterally extensive silicification, weathering to a conspicuous orange-brown horizon in the cliffs. Other distinctive markers include the 'Main Limestone' (Bed 49 of Trueman 1930), a 0.6 m limestone 5.5 m below the top of the unit, and the 'Sandwich Bed' (Bed 39 of Trueman 1930), a composite, 1.8 m limestone 10 m below the top (Plate 16). The unit ranges from the upper part of the *angulata* Zone into the *bucklandi* Zone. The lower part, up to the base of the 'Sandwich Bed', falls largely within the *conybeari* Subzone; at Nash Point [SS 9148 6834] these beds are locally rich in *Vermiceras*. The upper part of Unit B, including the 'Sandwich Bed', falls within the *rotiforme* Subzone, as indicated by the occurrence, at intervals, of *Coroniceras* sp. in the sequence at Nash Point (Trueman, 1930).

Unit C (beds 56 to 61 of Trueman 1930), up to 5 m thick, comprises massive, composite limestone beds with wispy mudstone partings (Plates 15 and 16) and is characterised by high limestone–mudstone ratios up to 85:15 (Figure 9). It differs from adjacent units in the coarseness of the skeletal debris in its consituent beds; many exhibit packstone textures and are rich in crinoidal debris and bone fragments, commonly forming coarse basal lags that fine upwards. The homogenising effects of bioturbation are evident, as in other units, but cross-lamination is locally preserved. The effects of silicification, including chert nodules and beekitised fossils, also commonly occur. The whole of the unit lies within the upper part of the *rotiforme* Subzone.

Unit D is easily eroded because of its high mudstone content, and tends to be preserved only in downfaulted structures at intervals along the coast; the lowest 12 m occur at Nash Point [SS 915 683], but a complete section is nowhere exposed. The removal of this unit down to the more massive limestones of Unit C appears to have been a major control on the level of inland erosion of the Lias plateau.

Unit D has a relatively low limestone–mudstone ratio of 60:40. The limestones, although planar-bedded locally, are generally nodular and the mudstones pinch and swell accordingly. A prominent 0.45 m planar limestone bed, informally termed the 'Upper Limestone', occurs 7 m above the base. The unit probably ranges from the upper part of the *bucklandi* Zone into the *semicostatum* Zone. The zonal boundary has not been determined exactly, but appears to lie within the lowermost beds; these beds at Nash Point have yielded *Coroniceras* cf. *caprotinum*, indicative of the upper *rotiforme* Subzone or the overlying *semicostatum* Zone.

The Porthkerry Formation of Dunraven Bay [SS 885 730] lies within a narrow graben between the Slade and Dunraven faults, which extends east-north-eastwards towards Cowbridge. On the northern side of the bay, at Dancing Stones [SS 8828 7322], a marginal facies, containing *Schlotheimia* sp., indicative of the *angulata* Zone, is overlain by 19 m of planar-bedded wackestones and calcareous mudstones in ratios of 60:40. The latter have yielded *Vermiceras* cf. *solaroides*, *V. conybeari* and *V. caesar* from the lowest 4 m, representing the *conybeari* Subzone, whilst the highest beds are tentatively assigned to the late *conybeari* or *rotiforme* Subzone from the occurrence of *Coroniceras* cf. *longidomus* in the upper 3 m. The faunal evidence therefore, suggests that the lower part of the Porthkerry Formation in Dunraven Bay is the equivalent of Unit B exposed elsewhere along the coast. The distinctive horizons of amalgamated limestone beds that characterise Unit B are not recognised within the Dunraven graben; this may reflect a change in the limestone to mudstone ratios, particularily in the upper part of the unit.

Overlying these strata in Dunraven Bay is a conspicuous unit, informally termed the 'Seamouth Limestone', comprising 4 m of fine- to medium-grained wackestone–packstones with thin mudstone partings. The whole of the unit is assigned to the *rotiforme* Subzone from the occurrence, at the top, of *Coroniceras* cf. *rotiforme*. Faunal and lithological evidence therefore, together with thickness considerations, suggests that the 'Seamouth Limestone' can be correlated with Unit C of the main coastal sections.

The 'Seamouth Limestone' is overlain by 8.5 m of nodular- and wavy-bedded limestones and mudstones in ratios of 45:55; these are succeeded by 3.7 m of planar-bedded limestones and mudstones in ratios of 60:40, overlain in turn by 15 m of mudstones with nodular limestones in a 50:50 ratio. The highest beds at Dunraven Bay are 10 m of planar-bedded limestones and mudstones in ratios of 75:25. The lower part of the sequence overlying the 'Seamouth Limestone' probably equates with Unit D of Nash Point, although bed-for-bed correlation is not possible. The lowermost beds fall within the *rotiforme* Subzone, as confirmed by the presence of *Coroniceras* cf. *rotiforme*. However, the occurrence of *Arnioceras* and *Euagassiceras* 6 m above the 'Seamouth Limestone' indicates that the higher beds can be tentatively assigned to the middle or upper part of the *semicostatum* Zone; this suggests that the *bucklandi–semicostatum* zonal boundary occurs within the lowermost part of the unit, similar to its position in Unit D at Nash Point.

The top of the Porthkerry Formation is not exposed in the Vale of Glamorgan. Coastal sections, including that of Dunraven, give a minimum total thickness of 104 m. Inland exposures are poor and it is difficult to subdivide the formation or to correlate with coastal sections. A borehole on the Bridgend Industrial Estate [SS 9225 7867] (Appendix 2), which proved 150 m of Blue Lias, penetrated up to 120 m of the formation, the youngest beds yielding *Arnioceras* cf. *semicostatum* at a depth of 14 m; the *bucklandi–semicostatum* zonal boundary was located at about 32 m depth. Excavations in Bridgend, notably at the Ford Motor Co. factory [SS 935 783], have yielded *Euagassiceras resupinatum* (Hodges, 1986), representing the highest (*resupinatum*) subzone of the *semicostatum* Zone. These beds probably correlate with the highest parts of the sequence at Dunraven Bay and are the youngest Blue Lias exposed in the district.

Localities: Continuous coastal section from Rhoose Point [ST 065 655] to Dunraven Bay [SS 885 730]; disused railway cutting, Bridgend [SS 912 800].

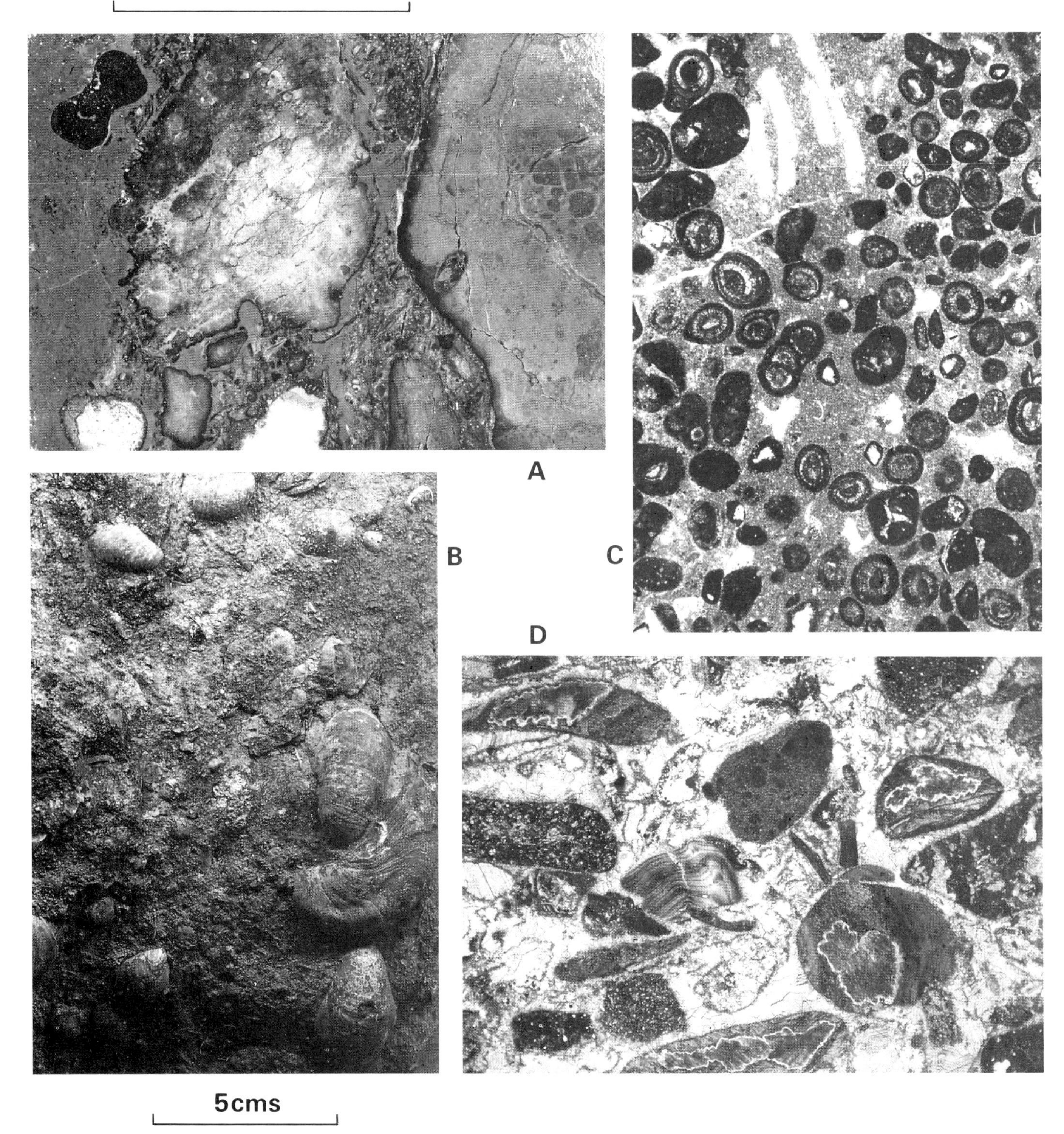

Plate 17 Lower Lias marginal facies: textural and lithological variations

A Conglomeratic lag, comprising bored and encrusted pebbles of reworked caliche carbonate with well-developed glaebular texture (Lilstock Formation, marginal facies), in a mixed matrix of bivalve coquina and calcite mudstone; basal Lower Lias, Bridgend Hospital borehole [SS 911 811]

B Calcarenite with numerous *Gryphaea*; Lower Lias marginal facies, near Colwinston [SS 9301 9514]

C Ooid packstone–wackestone; Lower Lias marginal facies, near Laleston [SS 8737 7918]. Field of view 6.5 mm (E 56499)

D Coarse calcarenite, with limestone lithoclasts and partly silicified skeletal debris; Lower Lias marginal facies near Witland [SS 8657 7927]. Field of view 7.0 mm (E 56487)

Lower Lias marginal facies (LLi/mf)

The limestones and mudstones of the Blue Lias pass laterally into a marginal facies, which rests unconformably on the older rocks. It generally fringes the Dinantian limestones of the breached and eroded Cardiff–Cowbridge Anticline, but in places overlies Triassic rocks and is locally interbedded with the Blue Lias (Figure 9).

The marginal facies was sustained around a series of islands that were extant until, at least, the upper part of the *bucklandi* Zone, and possibly into the *semicostatum* Zone (Hodges, 1986). The earliest deposits, at the base of the St Mary's Well Bay Formation in the north-west of the district, are thin, intermittent conglomeratic lags, containing mudstone flakes and pebbles of concretionary dolomite from the underlying Penarth Group marginal facies (Plate 17a); at Witland [SS 864 796] these deposits are markedly erosive into the older rocks. The marginal facies of the St Mary's Well Bay Formation occurs at St Mary Hill [SS 965 786] and between Llandough [SS 990 728] and Llantrithyd [ST 040 731] (Figure 9); similar deposits in the vicinity of Welsh St Donats [ST 028 760] may also be of this age (Trueman 1922). They comprise flaggy and rubbly bedded, coarse, bioclastic packstones, largely of *Liostrea* fragments, with scattered small limestone chips, micritic peloids and oolite wisps, in a matrix of skeletal packstone and wackestone.

The marginal facies of the Lavernock Shales, exposed on the coast between Whitmore Stairs [SS 898 713] and Temple Bay [SS 890 725], comprises thinly bedded limestone-chip conglomerates intercalated with shales containing abundant Dinantian lithoclasts and bioclastic debris.

The Porthkerry Formation passes into a marginal facies of grey and buff limestone-chip conglomerates and calcarenites containing chert pebbles, carbonised wood fragments, variable amounts of bioclastic material and micritic peloids (Plates 17b and d). Selective silicification and replacive nodular chert occurs in places. Many beds are thinly laminated or cross-laminated, and locally bioturbated; they are commonly separated by mudstone partings and display scoured surfaces. Interbedded oolitic and coquinoid packstones and grainstones occur at intervals (Plate 17c). Locally, within the Porthkerry Formation, the oolites form large lenticular bodies that exhibit shoaling-upwards motifs; in places, in the Cardiff district, their top surfaces are bored (Waters and Lawrence, 1987).

The 'Sutton Stone' and 'Southerndown Beds' (Tawney, 1866; Wobber, 1965, 1968; Hodges, 1986), exposed on the coast between Ogmore-by-Sea and Dunraven Bay, are local names given to distinctive types of marginal facies. The former is a sequence of massive and rubbly bedded, white, bioclastic limestones with corals, chert and Dinantian limestone fragments; the latter comprises blue-grey, well-bedded, limestone-chip conglomerates and calcarenites with shaly mudstone partings. These two types interdigitate in a complex fashion (Fletcher, 1988) and, although in the coastal sections the Sutton Stone appears to be a more proximal variety of the marginal facies (Wobber 1965), its relationship with the Southerndown Beds is not everywhere clear.

The diachronaiety of the marginal facies is well-illustrated in the vicinity of Llandough [SS 987 729], where it ranges, at outcrop, from the lateral equivalents of the St Mary's Well Bay Formation to the higher parts of the Porthkerry Formation. On the coast, between Ogmore-by-Sea and Trwyn-y-Witch, the earliest deposits, unconformably overlying Dinantian limestones (Appendix 1), fall within the *liasicus* Zone (cf. Trueman, 1922; Hodges, 1986) and are the correlatives of the Lavernock Shales. Their transition into the Lavernock Shales and Porthkerry Formation is exposed in coastal sections between Whitmore Stairs [SS 898 713] and Southerndown [SS 875 739], where they appear progressively higher in the sequence as the succession is traced northwards. At Whitmore Stairs they have replaced all but the upper 3.8 m of the Lavernock Shales and at Temple Bay [SS 890 725] they occupy all but the topmost 1.5 m of the formation. On the south side of Trwyn-y-Witch [SS 885 726], a marginal facies containing *Waehneroceras*, indicative of the *liasicus* Zone, has completely replaced the Lavernock Shales and the lowermost beds of the Porthkerry Formation. The marginal facies of the *angulata* Zone, exposed in Dunraven Bay [SS 883 732], is the lateral equivalent of Unit A and the lower part of Unit B of the Porthkerry Formation; it progressively replaces the higher beds until, on the northern side of the Bay, it occupies the whole of the sequence up to the 'Seamouth Limestone'. At Southerndown, the highest rocks of marginal facies fall within the *bucklandi* Zone (Trueman, 1922; Hallam, 1960).

Inland, at Castle-upon-Alun [SS 912 747], marginal facies limestones pass laterally into the Porthkerry Formation containing *Vermiceras* sp. of the *bucklandi* Zone; those near Wick [SS 923 723] also fall within this zone (Trueman, 1922). To the north-east of Stembridge [SS 946 742], rocks of the Porthkerry Formation containing *Arnioceras* cf. *semicostatum* are replaced by the marginal facies lying within the graben formed by the extension of the Slade and Dunraven faults. Lenses of oolitic and coarse bioclastic marginal facies limestones in the Bridgend area are the equivalents of strata ranging from the *rotiforme* Subzone of the *bucklandi* Zone to the *semicostatum* Zone (Trueman, 1922; Hodges, 1986).

The marginal facies generally represents a sequence of contemporaneous shoreface deposits developed during the late Triassic and Lower Jurassic transgression. Its absence from the eastern part of the district, where the passage from the Penarth Group into the Blue Lias is conformable, suggests that this area was inundated at an early stage. The transgressive conglomeratic lags at the base of the St Mary's Well Bay Formation in the north-west indicate local non-sequence. The marginal facies of the St Mary's Well Bay Formation first appears to the west of Llantrithyd [ST 040 731] and in the vicinity of Hensol Forest [ST 047 755]. These dominantly bioclastic limestones were probably former shell-banks that developed in active shoal areas over bathymetric highs (Figure 9) or adjacent to a subdued hinterland contributing little coarse sediment.

The marginal facies equivalents of the Lavernock Shales and Porthkerry Formation locally onlapped the Dinantian massifs across a series of planation surfaces exposed, notably, from Southerndown to Ogmore-by-Sea (Fletcher, 1988), on Tair Cross Down [SS 918 760] and Ewenny Down [SS 910 760]. Periodic fault activity was probably important along the margins of the massifs in maintaining the high-energy shorelines that sustained the marginal facies until, at least, *bucklandi* Zone times (Figure 9). Facies boundaries are broadly parallel to the line of several major fault zones, including the north–south plexus at Penllyn [SS 973 768] and the Slade and Dunraven faults. The latter were important controls on the position of the shoreline between Dunraven and Cowbridge, and probably influenced sedimentation within the Dunraven graben to the extent that bed-for-bed correlation with the Porthkerry Formation of the main coastal sections cannot be achieved.

Regional uplift along the Vale of Glamorgan Axis is indicated by a major shallowing event within the Porthkerry Formation. This event is reflected in the upward increase in the percentage of limestone and the thickness of the limestone beds, culminating, in the upper part of the *bucklandi* Zone, with the widespread progradation of the marginal facies into the offshore areas (Lawrence and Waters, 1978; Figure 9). Within the Porthkerry Formation, the acme of the event is marked by Unit C, with its reduced shale content, abundant coarse, winnowed bioclastic material, fining-upwards beds and cross-lamination, providing evidence of increased current activity. The lenticular bodies of marginal facies within the Porthkerry Formation in the Bridgend area also appear to have developed at this time, and are interpreted as localised shoals of oolitic and bioclastic material developed during regression, possibly related to uplift on underlying fault blocks.

Marine transgression late in *bucklandi* Zone times, continuing during the *semicostatum* Zone, terminated this shoaling episode and probably accomplished the final drowning of the district. Remnant islands may have briefly sustained a marginal facies during the early part of this period (Hodges, 1986), but their inundation allowed the regional deposition of Unit D of the Porthkerry Formation.

Localities: Disused quarry in Bridgend [SS 909 798]; cliff sections, Ogmore-by-Sea [SS 871 741]; *disused quarry, Tyn-y-Caeau [SS 9502 7878]; disused quarry, west of Llantrithyd [ST 0324 7303]; Longlands Quarry [SS 928 772].*

SEVEN

Quaternary

INTRODUCTION

The major climatic changes that occurred during the Quaternary resulted in the southward spread of ice over much of northern Europe. Several glacial events are known to have occurred, separated by interglacial or interstadial periods when the climate ameliorated. Two glaciations, with an intervening warm temperate interglacial period, have been recognised in South Wales (Bowen, 1973a, 1973b, 1974). The deposits of the earliest (pre-Ipswichian) glaciation occur south of the limits of the latest (Devensian) glaciation. The Ipswichian interglacial is recognised from raised beach deposits, mainly on the Gower Peninsula and in Pembrokeshire.

The glacial deposits of the Bridgend district comprise tills, morainic and fluvioglacial sand and gravel and laminated silts and clays. They were mostly deposited during the Devensian (the 'Newer Drift' of Charlesworth, 1929); however, the occurrence of supposed pre-Ipswichian till at Ewenny [SS 903 777] and Pencoed [SS 959 817] (Mitchell, 1960; Bowen, 1973a) has been cited as evidence of an earlier glaciation in the Vale of Glamorgan by ice of Irish Sea origin (Bowen, 1970, 1973a, 1974). There are no dated Ipswichian deposits within the district and evidence for the interglacial is largely circumstantial, being based on the recognition of supposed palaeosols (Crampton, 1964, 1966) of this age (Bowen, 1970, 1974).

Postglacial deposits and landforms range in age from the late Devensian, through the Flandrian, to recent times. They mainly comprise alluvial and head deposits, but also include the products of contemporaneous coastal processes and man-made features.

GLACIAL DEPOSITS

These deposits cover most of the north-eastern part of the district, are present within the main river valleys, and occur on the coastal tract between Kenfig [SS 790 820] and Porthcawl [SS 808 782] (Figure 2). The expanse of glacial drift in the north-east is generally regarded as part of an end-moraine of 'Newer Drift' (Charlesworth, 1929) deposited by a Glamorgan piedmont glacier; the glacial drift of Kenfig is envisaged as being deposited on the flanks of a second piedmont glacier which debouched into Swansea Bay (Charlesworth, 1929; Bowen, 1970). The glacial deposits of the lower Ewenny valley, hitherto regarded as remnants of an earlier (pre-Ipswichian) glaciation (Strahan and Cantrill, 1904; Mitchell, 1960; Bowen, 1973a, 1973b), occur south of the limits of the Devensian ice.

The glacial deposits generally consist of gravelly tills, containing lenses of sand and gravel. Fluvioglacial outwash gravels occur within, and beyond, the limits of the former Devensian ice-front, forming terraces in places and underlying the alluvium of the main valleys. Laminated silts and clays occur as lenses within the tills and outwash gravels.

Till

Till deposits range from generally structureless, stiff, very stony, sandy and silty clays (boulder clays) to clayey gravels. They are generally mottled grey and yellow, but incorporated Triassic and Old Red Sandstone material commonly imparts a reddish hue, and till in contact with such bedrock is often bright red. Pebbles and cobbles within the tills are dominantly of Silesian sandstones, the breakdown of which, promoted by acid soil conditions, commonly gives rise to a weathered profile of buff, clayey sand. Lenses of contemporaneous sand and gravel occur locally.

The till deposits are extremely variable in thickness, commonly forming a characteristic hummocky topography of irregular mounds and ridges with intervening sinuous depressions and kettle holes. These landforms are generally accepted as representing an extensive ablation terrain, resulting from the melt-out of stagnant ice.

Localities: Llanilid Opencast Site [SS 986 820]; Garwa Farm Quarry [SS 981 797].

Glacial sand and gravel

These deposits range from poorly sorted clayey sands and gravels to moderately well-sorted, imbricated, pebble-cobble gravels with a coarse-grained sand matrix. Thin beds of laminated and cross-laminated, pebbly sands occur at intervals, with lenses of silt and clay less commonly. Bedding within the sands and gravels varies from well ordered to chaotic. Poorly sorted deposits locally grade to sandy and gravelly till.

The sands and gravels are all of Devensian age and record a variety of fluvioglacial environments, the majority being outwash deposits. They are preserved locally as dissected and generally modified terrace-like features in the main river valleys. In the Ogmore valley, between Bridgend and Ewenny, they form a well-defined terrace at 25 m above OD which, at its southern end, overlies and, in places, is incised into supposed pre-Ipswichian till deposits. Further small terrace remnants occur at Ewenny Priory [SS 912 778], in the Thaw valley between Howe Mill Farm [ST 006 725] and Gigman Mill [ST 016 714], and at Cowbridge [SS 993 747].

Locality: River Ogmore [SS 8613 7562].

Glacial silts and clays

These deposits comprise dark brown and purplish red clays and buff silts, with subordinate fine-grained sands and sporadic thin gravel lenses. They vary from well bedded and thinly laminated, to structureless. Small pebbles, scattered throughout, locally disturb the laminae and may represent 'dropstones' from ice melt-out; they are dominantly of sandstone, but also include coal and weathered chert. Horizons of disrupted and contorted laminae are probably collapse effects induced by ice meltout, or result from disturbance by overriding ice. The silts and clays are generally unfossiliferous, but a borehole at Glanwenny [SS 9030 7814] (Appendix 2) proved 4.35 m of reddish brown and purple silty clays with a sparse foraminiferal assemblage including the benthonic *Ammonia batavus* and *Protelphidium anghium*.

The glacial silts and clays probably accumulated in temporary lakes, locally impounded by morainic deposits and topographical obstacles, or in fluvioglacial environments. They occur in all the main valleys, where they commonly underlie alluvium or outwash gravels. In a former brickpit at Pencoed, a borehole [SS 9580 8216] (Appendix 2) in a deposit hitherto regarded as pre-Ipswichian till (Bowen, 1970, 1974) proved up to 17 m of barren, fine-grained laminated sands, silts and clays; these are locally overlain and disrupted by till of the Devensian ice-sheet. At Ewenny [SS 904 778], on the interfluve of the Ogmore and Ewenny valleys, glacial silts and clays are intercalated with gravels and gravelly tills, the latter having been considered pre-Ipswichian in age (Bowen, 1970, 1973a, 1974). Deposits of silt and clay also underlie fluvioglacial outwash gravels in Bridgend [SS 910 805] and have been proved in boreholes at Pontyclun [ST 0344 8191] (Appendix 2) and Pyle [SS 8391 8194].

Locality: Degraded pit at Pencoed [SS 957 821].

Buried valleys

The bedrock profiles of all the major valleys extend below OD in their lowermost parts (Anderson 1974), and all are filled with glacial and postglacial materials. The upper parts of the valleys within the Devensian ice limits contain thick sequences of till, laminated silts and clays and fluvioglacial sands and gravels beneath alluvial deposits. To the south of the Devensian ice front, beneath the alluvium of the major valleys, are sands and gravels with interbedded peats in places.

Downcutting was probably enhanced by a reduced sea-level during the Devensian glacial maximum, and ice may have further modified the valley profiles. Erosion by proglacial meltwaters is probably responsible for the steep-sided and V-shaped subdrift profile of the Ogmore valley in Bridgend (Wilson and Smith, 1984). The Ewenny valley between Ewenny and Pencoed has shallower sides with an irregular longitudinal profile. At Pencoed, a northerly-trending asymmetric rock basin is developed, the southern lip of which lies close to the probable position of the former ice front, at a point where the valley narrows between steep escarpments of Triassic rocks. Boreholes have proved deposits of till in the eastern part of the basin, whereas those penetrating the deeper, western parts show sequences of outwash gravels underlain by thick deposits of laminated silts and clays. The overdeepening and scouring of the basin may have resulted initially from erosion by subglacial meltwater under hydrostatic pressure, at the Devensian ice front. Meltwater from the ice front spilled into the lower Ewenny valley, but this outlet was probably sealed temporarily by ice or morainic deposits, allowing silts and clays to accumulate within the basin.

Correlation of glacial sequences

The earliest glacial deposits in South Wales are undated, but preceded the Ipswichian interglacial. They are preserved at several localities (Bowen, 1973a, 1973b, 1974), and it is generally accepted that they were derived from Irish Sea ice, which entered the Bristol Channel and impinged on the present coastline.

Glacial deposits at Pencoed and Ewenny have been regarded as till of Irish Sea derivation (Mitchell, 1960; Bowen, 1970, 1973a, 1974) from the reputed occurrence of shelly material and igneous erratics at the former locality (Howard and Small, 1901; Strahan and Cantrill, 1904). These deposits have also been cited as evidence that the Vale of Glamorgan was glaciated during the pre-Ipswichian period (termed the Pencoed Cold Stage by Bowen, 1970). The Pencoed deposit, however, has none of the characteristics of a till, nor is there any indication that it is older than the Devensian, the laminated sands, silts and clays revealing evidence of deposition in a temporarily impounded lake that developed at the margin of the Devensian ice-sheet.

The Ewenny deposit is a complex of till, gravels, silts and clays. The till contains abundant clasts of Silesian sandstones and appreciable amounts of weathered chert, probably derived from Dinantian or Jurassic limestones, but the occurrence of erratics of undoubted western provenance (Strahan and Cantrill, 1904) has not been verified. The till probably impounded ephemeral lakes, in which the silts and clays were deposited. The species of foraminifera within the Ewenny silts and clays are indicators of a temperate, marine environment, but are probably derived, with the Bristol Channel as the most obvious source. They may have been transported and redeposited by westerly derived ice, or introduced by aeolian reworking of coastal flats exposed during a glacial low-stand in sea-level; in the latter case, the foraminifera give no indication of the source of the ice that deposited the Ewenny till. The origin of this till is, therefore, equivocal; it may have been derived from the west during a pre-Ipswichian glaciation (Mitchell, 1960; Bowen, 1970, 1973a, 1974), but equally may have been deposited during the Devensian, either by ice from the north-east that penetrated the lower Ewenny valley ahead of the main ice front, or from westerly derived Devensian ice (Woodland and Evans, 1964).

There are no further deposits within the district that can be ascribed to the pre-Ipswichian glaciation, although circumstantial evidence has previously been cited to suggest that the Vale of Glamorgan was covered by ice during this period. It has been suggested that exotic heavy mineral suites within soil profiles indicated contamination from a relic cover of glacial material of Irish Sea derivation (Crampton, 1960, 1961), but they are more likely to have been introduced by aeolian action. Cobbles of Lower Lias limestone, within soil profiles overlying Lower Lias bedrock, were formerly thought to be of glacial origin (Crampton, 1966), but this is unlikely because they are generally derived by in-situ weathering of Lower Lias bedrock. The occurrence of westerly derived erratics in the Ely valley and as far east as Cardiff (Griffiths, 1939) has not been confirmed (Waters and Lawrence, 1987), there being a general absence of erratics south of the Devensian ice front.

There are no proven Ipswichian deposits in the Bridgend district, but the *terra rossa* and *terra fusca* affinities of local soil profiles have been interpreted as evidence of a former warm temperate climate (Crampton, 1964, 1966), and Bowen (1970, 1974) has suggested that they are palaeosols of Ipswichian age. There is, however, no evidence to suggest that these are anything other than modern soil profiles. The '*terra rossa*' soils on Newton Down [SS 838 795] are probably due to the weathering of Triassic bedrock or palaeokarst horizons within the Oxwich Head Limestone, both of which commonly impart a characteristic reddening to soil profiles. The shallow, free-draining '*terra fusca*' soils on parts of the Lower Lias outcrop generally reflect the limestone to mudstone ratios of the bedrock.

It has been suggested (Trenhaile, 1971) that ledges along the coast, between low tide and 12 m above OD, are remnant interglacial shore-platforms. These occur at Sker Point [SS 786 796], Porthcawl [SS 812 766] and Black Rocks [SS 867 742], but no deposits have been found on them to provide evidence of their age and, in general, the features appear to be the result of differential weathering of the Carboniferous and Triassic rocks.

The Devensian ice front in the district is broadly similar to that previously described (Charlesworth, 1929; Bowen, 1981; Figure 2). In the north-east, ice from the Glamorgan piedmont glacier debouched onto the low ground south of the coalfield, impinging on, and locally overriding, the Dinantian limestone escarpment on the southern limb of the Cardiff–Cowbridge Anticline. Devensian ice encroached onto the low ground north-west of Coity [SS 918 820], but north of Bridgend was held back by the major escarpment forming the southern margin of the coalfield. The tills, sands and gravels in the north-west of the district have been interpreted as a complex kamiform morainic belt deposited by Swansea Bay ice during the Margam Stage of the Devensian glaciation (Bowen, 1970).

POSTGLACIAL DEPOSITS

A wide range of deposits accumulated during the late Devensian climatic amelioration and Flandrian rise in sea-level. Peat formed in hollows within the extensive ablation terrain left by the retreating ice, and the change in base-level allowed alluvial deposits to aggrade in the major valleys. Head deposits, although conventionally regarded as a product of periglacial environments, have probably all accumulated since the Flandrian. In places, calcareous tufa has been deposited around springlines. Recent coastal deposits include estuarine alluvium, marine and storm gravel beach deposits and blown sand. Made ground and worked-out ground are man-made features commonly found near centres of population. Landslips occur in places on steep valley slopes, and coastal erosion is particularily active locally.

Alluvium and alluvial fans

Alluvium forms wide, flat tracts in the Ogmore, Ewenny, Thaw and Ely valleys. It generally consists of brown and grey silty clays, sands and gravels with peat horizons in places. Alluvial fans of silt and clay with intercalated calcareous tufa have formed at the confluence of tributary streams with the River Waycock, east of Penmark [ST 058 688].

River terrace deposits (undifferentiated)

River terrace deposits occur in the Ely valley north of Pontyclun [ST 037 817] and in the Ewenny valley east of Pencoed. They generally form flat expanses of coarse, poorly sorted gravels and ochreous brown clays, 3 to 5 m above the alluvium. Their age is uncertain, but they are probably no older than the late Devensian.

Peat

Most of the kettle holes and enclosed drainage systems on the till surface contain peat, silt and clay, locally with thin gravel lenses; the largest of these deposits is near Ystradowen [ST 025 787]. Peat, up to 2.5 m thick, has also accumulated in places within alluvial deposits.

Head

Head deposits, of varying thickness, have accumulated on the lower slopes and bottoms of many valleys. They generally comprise brown, silty and sandy clays with a variable pebble content, locally crudely graded and imbricated, the composition of which largely reflects the nature of the parent material upslope. Head deposits have formed by a variety of gravity-induced mass-movement processes including solifluction, soil-creep and downwash. The head deposits of Cwm Marcross [SS 915 684] and Cwm Tresilian [SS 946 684] are disposed in a series of terraces, which record the progressive lowering of erosive base-level and incision, with the formation of nick-points, by the streams that occupy these cwms. These events may reflect actual or relative falls in sea level during the Flandrian (Williams and others, 1981), but could equally have been formed in response to cliff retreat during historical times.

Calcareous tufa

This is a soft, cellular deposit of calcium carbonate, which has usually formed around organic debris such as twigs and leaves. It is commonly deposited by springs issuing from the Blue Lias, and has often been redistributed as slopewash, intercalated with head or alluvial deposits. At Cwm Nash [SS 905 700] pale brown tufa, interbedded with head deposits and containing interlayered soil horizons, has yielded a molluscan fauna (Bowen, 1970; Perkins and others, 1979) which records a series of climatic changes throughout the Flandrian.

Plate 18 Hanging valley resulting from erosion and retreat of cliffline of Lower Lias limestones and mudstones; Cwm Mawr [SS 8925 7227] (A 14349)

Estuarine alluvium

Deposits of estuarine alluvium, generally consisting of grey and brown clays and silts, are present in the tidal reaches of the Ogmore and Thaw valleys.

Marine beach and storm gravel beach deposits

Marine beach deposits generally comprise well-sorted sands and gravels occurring below high water mark. Storm gravel beach deposits, up to 5 m high, occur at intervals along the coast and consist of cobbles and boulders of Blue Lias.

Blown sand

Extensive dune fields, of fine-to medium-grained quartz sands, blanket the coastline at Kenfig Burrows [SS 790 815], Merthyr Mawr Warren [SS 860 771] and Ogmore Down [SS 882 760], with smaller deposits in Limpert Bay [ST 019 664]. The dunes are relatively stable at present, but major advances have occurred in historical times, the most recent being that of the sixteenth century which buried the town of Old Kenfig [SS 795 822].

Made ground and worked-out ground

Made ground comprises tipped deposits of variable composition and thickness, such as colliery spoil heaps or rubbish tips, which generally form well-defined features, but are often unrecognisable when regraded or landscaped. Worked-out ground is generally restricted to sites of opencast coal and ironstone extraction backfilled with a range of materials; when such sites are restored or regraded it is often difficult to determine the limits of excavation.

Landslips

Landslips result from gravity-induced failure of unstable slopes, generally assisted by high fluid pressures in the slipped material. They are mainly associated with prominent springlines, particularily along the junction of the Lavernock Shales and Porthkerry Formation in deep valleys around Llancarfan [ST 052 704].

Coastal erosion

Erosion of the Liassic cliffs has continued throughout postglacial times. Evidence of cliff-line retreat can be observed from the part-eroded early Iron Age fortifications, the hanging valleys of Cwm Mawr [SS 893 723] (Plate 18) and Cwm Bach [SS 897 719], and the numerous toppled blocks that litter the foreshore and wave-cut platform.

EIGHT

Structure and mineralisation

INTRODUCTION

Several major tectonic episodes have affected South Wales since the Lower Devonian. The earliest event recognised within the district is the intra-Devonian, late Caledonian regional uplift, resulting in erosion, widespread unconformity and the absence of Middle Devonian sediments. Intra-Carboniferous movements on major structures preceded the main Variscan deformation and influenced sedimentation from early Dinantian to late Silesian times. The Variscan deformation of South Wales occurred towards the end of the Carboniferous. In the Bridgend district, it gave rise to a major east–west fold, the Cardiff–Cowbridge Anticline, and a series of thrusts and 'cross-faults'. The tectonic controls that affected Mesozoic sedimentation within the district probably included extensional reactivation of the Variscan thrusts and 'cross-faults' as normal faults. Subsequent movements were generally confined to these structures and postdate the Lower Jurassic (Figure 1).

The geological history of the district records the repeated influence of a major east–west structural hinge-line underlying the Vale of Glamorgan. This hinge-line, termed the Vale of Glamorgan Axis (Waters and Lawrence, 1987), is probably related to a concealed basement fault and is broadly coincident with the axial trace of the Cardiff–Cowbridge Anticline, a structure developed partly by inversion of the fault.

INTRA-DEVONIAN MOVEMENTS

Evidence of the early tectonic history of the district is incomplete, but during the Lower Palaeozoic and for much of the Upper Palaeozoic it was part of a relatively stable cratonic platform, alternately located on the margins of the Lower Palaeozoic Welsh Basin and the Upper Palaeozoic basin of south-west England. The earliest recorded tectonic events are mid-Devonian, related to the final deformation and uplift of the paratectonic Welsh Caledonides, but regional instability to the south during the Lower Devonian may have been responsible for the introduction into the district of the Llanishen Conglomerate; the idea of a 'Bristol Channel Landmass' (Anderton and others, 1979), possibly a complex fault zone with elements of strike-slip (Tunbridge, 1986), has been proposed to explain these southerly derived sediments. The late Caledonian regional uplift resulted in major unconformity and the non deposition of Middle Devonian sediments. The marked southward overstep of the Upper Devonian across the Lower Devonian Brownstones in the district provides the earliest evidence of activity on the Vale of Glamorgan Axis. Upper Devonian movements are recorded in the pronounced southward thickening of the Cwrt-yr-ala Formation across the Axis.

INTRA-CARBONIFEROUS MOVEMENTS

Dinantian movements on the Vale of Glamorgan Axis were probably generated by extensional processes, involving the formation of the Upper Palaeozoic basin of south-west England (Leeder, 1976, 1982; Bott, 1984). In the Bridgend district these movements were largely responsible for the major thickness and facies variations in the lower part of the Dinantian, occurring across the hinge of the later Cardiff–Cowbridge Anticline. The marked southward thickening of the Courceyan and Chadian sequences, particularily the Friars Point Limestone and Gully Oolite (Figure 5), suggests that the Axis behaved as a concealed syndepositional fault, producing downwarp to the south; continued downwarp during the late Chadian regression is indicated by the absence of karstic and pedogenic effects at the top of the Gully Oolite in the south of the district. The Vale of Glamorgan Axis continued to influence Dinantian sedimentation into the Arundian, controlling the initiation of the barrier facies of the High Tor Limestone which, in turn, restricted the southward extent of the Caswell Bay Mudstone peritidal sequences (Figure 5).

The waning influence of the Vale of Glamorgan Axis is apparent from Arundian times. The late Arundian and Holkerian sediments of the district record an east–west facies differentiation, related to uplift along the Usk Anticline to the east. Continued uplift on this structure resulted in the progressive eastward overstep of the Namurian sequence (George, 1956; Figure 6). Instability during the Westphalian produced marked local thickness variations across the Miskin Fault, and probably generated the synsedimentary growth faults that characterise parts of the sequence in the north of the district (Elliot and Ladipo, 1981). These Namurian and Westphalian movements have been attributed to basement controls, resolving the principal Variscan stresses (Walmsley, 1959; Squirrell and Tucker, 1960; Owen and Weaver, 1983), and to strike-slip reactivation of major basement lineaments, producing localised deformation and uplift of the Variscan foreland (Williams and Chapman, 1986).

VARISCAN STRUCTURES

The main Variscan deformation in Wales occurred during the late Silesian. Various criteria have been used to define the position of a Variscan orogenic front (see Matthews, 1974; Dunning, 1977); some authors (e.g. Freshney and Taylor, 1980) place it across the Vale of Glamorgan, but other suggested positions are in the Bristol Channel (Owen, 1974; Matthews, 1974) and north of the district as a major decollement surface under the coalfield (Shackleton and others, 1982; Shackleton, 1984). However, it has been suggested that the 'front' is a diffuse zone across which deformation dies out gradually (Williams and Chapman, 1986), and that it occurs significantly farther north than previously proposed.

The Variscan deformation in the district is represented by a major fold, the Cardiff–Cowbridge Anticline, with several second-order folds, a series of east–west thrusts and a number of 'cross-faults' with general north–south or north-west–south-east orientations.

Folds

The Cardiff–Cowbridge Anticline is an open, apparently symmetrical, upright structure with an east–west axial trace, a horizontal or gentle westerly plunging fold axis, and fold limbs which dip, on average, between 25 and 35°. The fold profile however, is more complex, with thrusts on both limbs, which locally transect second-order, medium-scale, east–west anticlines on Stalling Down [ST 016 750] and in Hensol Park [ST 040 787] (Figure 10); a further medium-scale fold occurs in the hinge zone of the anticline at Candleston [SS 872 773] (George, 1933).

The axial trace of the Cardiff–Cowbridge Anticline and the Vale of Glamorgan Axis are broadly coincident, a fact that is reflected in the major thickness and facies changes within the lower part of the Dinantian, between the northern and southern limbs of the fold. Their relationship suggests that reactivation of the Vale of Glamorgan Axis, probably as a major Variscan thrust within basement rocks, was an important control on the growth of the anticline. It is unlikely that the Cardiff–Cowbridge Anticline is a westward continuation of the Usk Anticline as previously suggested (George, 1956; Owen, 1971, 1974) in view of the discordance and differing history of movement between the latter structure and the Vale of Glamorgan Axis (Waters and Lawrence, 1987).

Thrust-faults

Thrusting appears to be a significant feature of Variscan deformation in the district as elsewhere in South Wales (e.g. Hancock and others, 1983). Thrust-faults are exposed in Pant-y-ffynnon Quarry [ST 046 740] and Ruthin Quarry [SS 974 793], and have also been proved by boreholes and mineworkings in the north. The majority, however, are unexposed and in some cases are partially concealed or inferred to lie beneath younger rocks. Although several of these faults have undergone subsequent reactivation, including reversals of movement, residual shortening estimates of 15 to 20 per cent are indicated in cross-sections of the Cardiff–Cowbridge Anticline (Figure 10).

The majority of thrusts are oriented approximately east–west, and occur on both limbs of the Cardiff–Cowbridge Anticline (Figure 10). Those on the southern limb are mainly directed north and disposed as a series of relatively high-angle ramps, cutting up section to the north. The thrusts on the northern limb are generally directed south, and are probably related to similarly oriented structures occurring in a complex zone of folding and thrusting between Cefnyrhendy [ST 045 820] and Llanilid [SS 980 825] and forming part of a broad zone of disturbance that affects the Silesian sequences on the southern margins of the coalfield. The major thrust in this area is the Llanharan Thrust, downthrowing about 220 m to the south, but there are numerous smaller thrusts as well as evidence of bedding-parallel slip along coal seams. The thrusts commonly transect the limbs of a series of tight, southward-verging, medium-scale anticlines and synclines but, in places, are coincident with the fold axial surfaces. It has previously been suggested (Woodland and Evans, 1964) that these folds were generated by drag, due to northern transport of the thrust footwalls, which were accordingly termed 'underthrusts'; however, all these structures may be features of back-thrusting from a lower decollement.

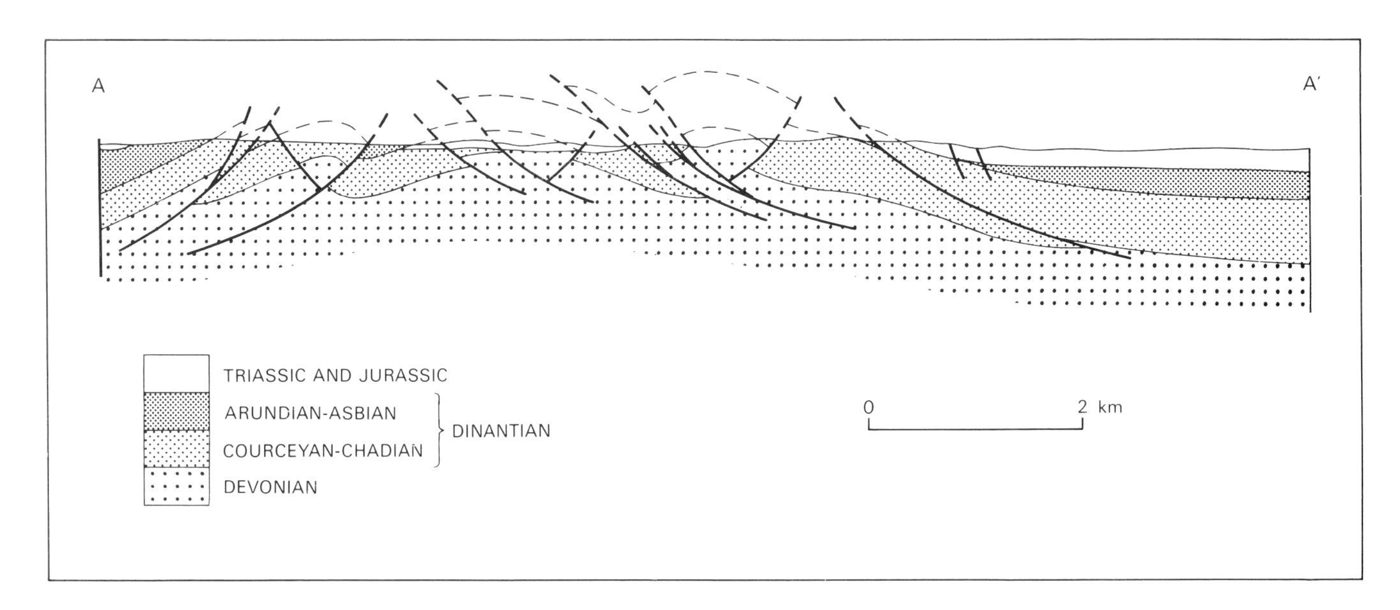

Figure 10 Structural cross-section of the Cardiff–Cowbridge Anticline along line A–A′ on Figure 1 (vertical = horizontal scale).

Plate 19 Large monoclinal fold with associated faulting; Porthkerry Formation, Stout Point [SS 9736 6700] (A 14324)

Cross-faults

These are a series of north-north-westerly and north-north-easterly trending faults transecting the Silesian in the north of the district and common on the northern limb of the Cardiff–Cowbridge Anticline. Faults with similar trends, associated with the Newton Fault complex, occur in the vicinity of Porthcawl (Figure 1). Several faults are continuations of similarly oriented 'cross-faults', previously recorded in the Pontypridd district (Woodland and Evans, 1964), which appear to largely postdate the Variscan thrusting but, in some cases, were broadly synchronous with fold development within the coalfield. Triassic rocks unconformably overlie a number of these faults in the Bridgend district, demonstrating the probability that they are Variscan structures, but many have subsequently been reactivated in Mesozoic, and possibly Tertiary times (Anderson and Owen, 1968; Gayer and Criddle, 1969); others, such as the Miskin Fault, which now has a westerly downthrow of 400 m, provide evidence of pre-Variscan instability.

MESOZOIC AND TERTIARY MOVEMENTS

Considerable uplift marked the final phase of the Variscan Orogeny. During the Permo-Triassic the district was situated on the northern margin of a major east–west, fault-bounded extensional basin initiated in the Bristol Channel–Somerset region (Whittaker, 1975). There is some evidence that, during the Triassic, the Palaeozoic massifs were regenerated in places by active faults along their margins (Bluck, 1965; Waters and Lawrence, 1987). A number of faults were clearly active during the Lower Jurassic, sustaining the shorelines against which the Lower Lias marginal facies accumulated (Figure 1); these include reactivated Variscan thrusts such as the Slade and Dunraven faults, and the Penllyn fault plexus. Other synsedimentary faults may be concealed by Lower Lias onlap; near Llangan [SS 955 772], a concealed fault is suspected where a mineshaft penetrated 45 m of the marginal facies close to its unconformity with the Dinantian. This Lower Jurassic fault activity is attributed to renewed activity on the Vale of Glamorgan Axis, which produced major regression in the district during the *bucklandi* Zone. Uplift along the Axis was probably compensated by subsidence to the south in the Bristol Channel Basin; similar differential movements are recorded across the Mendip Axis at this time (Sellwood and Jenkins, 1975).

Post-Lower Jurassic reactivation has taken place on most of the faults, resulting in different amounts of Palaeozoic and Mesozoic displacement and, in places, producing reversals of throw. Sinistral strike-slip movements on the Dunraven Fault, probably of limited extent, are indicated in coastal exposures by low-angle slickensides, calcite-filled tensional vein arrays and small, anticlockwise-transected folds. Faulting, at depth, has locally produced small- to medium-scale monoclinal drapes at higher levels within the Blue Lias (Plate 19), one of which trends north-north-east between

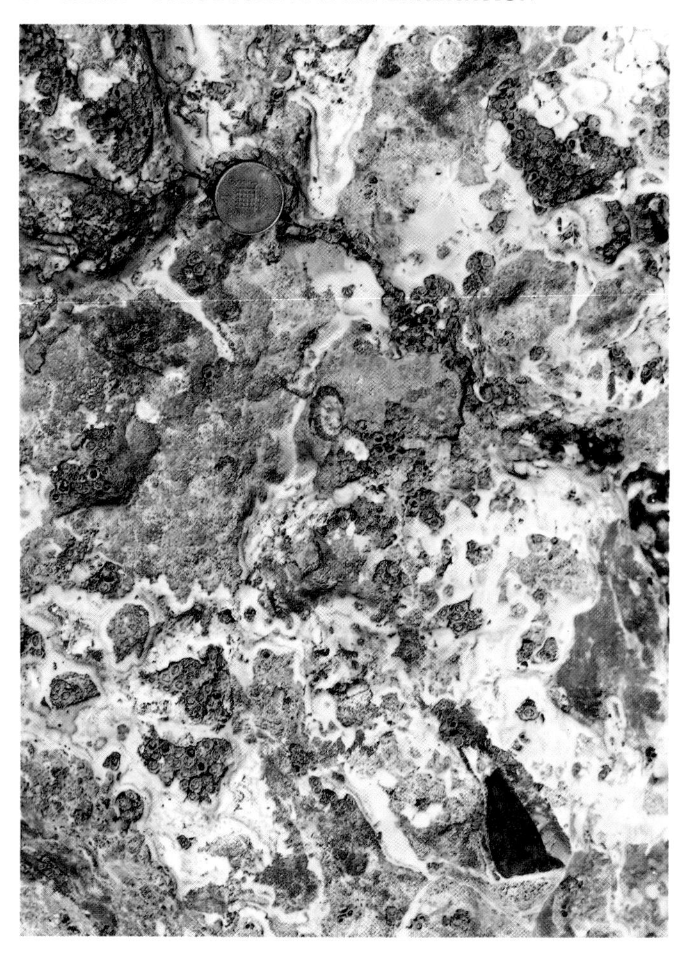

Plate 20 Barytes-galena mineralisation; Lower Lias marginal facies, Ogmore-by-Sea [SS 8725 7403]

Barytes locally replaces coral and bivalve debris and occurs in two forms, a pale, cryptocrystalline variety and a white, crystalline type. Small cubes of blue-grey galena occur within the barytes (A 14335)

Penmark [ST 058 689] and Nant Whitton [ST 070 721]. The age of these latest movements are unknown, but Miocene displacements, in response to Alpine tectonics, have been suggested for the north-north-westerly 'cross-faults' (Anderson and Owen, 1968) and the east–west faults (Gayer and Criddle, 1969).

MINERALISATION

Mineralisation occurs in two distinct forms within the district. The first is dominated by lead (galena), barytes and calcite, with minor amounts of iron, zinc and copper minerals, and the second is dominantly of iron (hematite) and silica (quartz).

Lead mineralisation

Lead mineralisation occurs in places within the Dinantian limestones and has also affected the Triassic and Jurassic marginal facies rocks. The bulk of the mineralised material consists of calcite and barytes, with galena commonly occurring in disseminated form but, locally, as thin stringers, or in pockets up to 5 cm in diameter (Plate 20). Minor amounts of chalcopyrite and sphalerite are present, in places associated with malachite, azurite and rosasite (Fletcher and Young, 1987) as secondary alteration products. In the Dinantian, the mineralisation takes the form of narrow vein systems, invading faults and major joint planes and locally disrupting the Triassic fissure-fills. In the marginal facies, in a narrow, irregular zone immediately above the unconformity with the Dinantian, it acts as a cement to the conglomerates and breccias, filling voids and, locally, replacing fossils (Plate 20). In places, at Ogmore-by-Sea, mineralised marginal facies conglomerates fill small veins within the underlying Dinantian limestone.

The mineralisation is probably of hydrothermal origin, but its source is unknown. The veins in the Dinantian appear to have acted as conduits to exhalative deposits, which invaded semilithified marginal facies conglomerates and breccias and spread onto the Jurassic sea floor. An Hettangian or Sinemurian age is indicated for the bulk of the mineralisation, but the possibility of an earlier hydrothermal event, with subsequent Jurassic remobilisation, cannot be excluded (Fletcher, 1988).

Iron (hematite) mineralisation

This form of mineralisation occurs in ore bodies confined to the north-east of the district and extensively worked in a broad zone extending east-north-eastwards from Llanharry [ST 005 805]. It mainly comprises colloidal and crystalline hematite, commonly altered to goethite, with quartz, calcite and dolomite; pyrolusite has been recorded at Trecastell [ST 017 812] (Strahan and Cantrill, 1904). Pyrite occurs in minor amounts, mainly within the palaeosol horizons of the Oxwich Head Limestone, where it is partially replaced by hematite (Williams, 1958). The hematite ore fills large cavities and fissures in Dinantian limestones, generally beneath the Triassic unconformity, at their contact with impermeable Namurian mudstones. The ore bodies occur along the line of the major north-north-westerly 'cross-faults', usually on the downthrow side; they thin appreciably at depth, locally becoming strata-bound in a zone to the side of the fault.

The origin of the ores has been ascribed either to a hydrothermal source or to circulation of meteoric groundwaters, leaching iron and silica from adjacent Silesian or Triassic strata. Textural studies (Williams, 1958; Gayer and Criddle, 1973) have indicated that the ores formed by a combination of replacement and cavity-infill processes, the hematite substituting the dolomitised Dinantian limestones and an earlier sulphide (pyrite) phase. The most likely explanation therefore involves a combination of these processes, with an earlier hydrothermal (sulphide) phase being followed by a replacive phase of meteoric circulation.

NINE

Economic geology

INTRODUCTION

Quarrying of limestone, for aggregate and cement, and opencast coal extraction are, currently, the two major resource activities of the district. The mining of iron ore was previously important and lead ore has been worked sporadically. Minor resources have, in the past, included sandstones and sands and gravels, mainly for aggregate and building purposes, and clays for pottery. Water resources include a number of potential aquifers.

LIMESTONES

The Dinantian limestones support a quarrying industry with an annual output in 1984 of about 2.4 million tonnes of limestone products, mostly for aggregate, but also for cement making and chemical-grade limestone (Table 1). The limestones occur in thick, uniform beds which are easily quarried, the only problem being the fissure and joint systems and numerous swallow-holes, many of which are filled with poorly consolidated Triassic deposits. The physical and mechanical properties of the Dinantian limestones render them good-quality aggregate material (Harrison, 1984); their widespread use as aggregate results from a shortage of local gravel resources. The purest limestones, of chemical grade quality, are the oolitic formations, notably the Cornelly and Gully Oolites, and high-purity limestones also occur in the Stormy Limestone Formation and Oxwich Head Limestone. The dolomitised Friars Point Limestone generally has a lower MgO ratio than that required of commercial dolomites for refractory material or as a fluxing agent.

The Blue Lias (Porthkerry Formation) is at present being quarried for cement near Aberthaw (Table 2). The main impurities that affect cement making are present in only small quantities within the Porthkerry Formation and its potential as a cement-grade resource extends to the whole area of its outcrop. The main factor affecting its quality is the varying limestone to mudstone ratios throughout the formation. At present the low calcium carbonate content of the stone at Aberthaw is augmented by the addition of Dinantian limestone to provide a limestone to mudstone ratio of 3:1, an ideal mix for cement manufacture.

COAL

Coal has been mined in South Wales from an early date, certainly as far back as the 16th century. The main expansion in mining occurred during the early 19th century and extraction reached a peak in the early 20th century, since when it has slowly declined. Of the four mines within the district, Llanharan Colliery [SS 9960 8230] was the largest and most

Table 1 Major quarries in Dinantian limestones and their products

Quarry	Locality	Principal quarried formations	Products
STORMY WEST	South Cornelly SS 846 806	Stormy Lst./Oxwich Head Lst.	Aggregates
STORMY DOWN	South Cornelly SS 842 804	Cornelly Oolite/ Stormy Lst.	Roadstone, concrete aggregate
CORNELLY	South Cornelly SS 835 800	Cornelly Oolite/ Stormy Lst.	Roadstone, concrete aggregate, steel flux
GAEN	South Cornelly SS 824 804	Stormy Lst./Oxwich Head Lst.	Aggregates
GROVE	South Cornelly SS 822 797	Cornelly Oolite/ Stormy Lst./ Oxwich Head Lst.	Aggregates
PONTALUN	St Brides Major SS 897 765	Gully Oolite/High Tor Lst.	Chemical grade limestone, aggregate
PANT	St Brides Major SS 897 760	Gully Oolite/High Tor Lst.	Roadstone, concrete aggregate
RUTHIN	Pencoed SS 975 795	Cornelly Oolite/ Stormy Lst./ Oxwich Head Lst.	Cement, roadstone, concrete aggregate
ARGOED ISHA	Llanharry SS 993 790	High Tor Lst./ Cornelly Oolite	Roadstone, concrete aggregate
FOREST WOOD	Llanharry ST 015 797	Gully Oolite/Caswell Bay Mudstone/ High Tor Lst.	Roadstone, concrete aggregate, blockstone
PANT-Y-FFYNNON	Bonvilston ST 045 740	Friars Point Lst.	Aggregates
HENDY	Miskin ST 054 810	Friars Point Lst./ Gully Oolite	Aggregates

Table 2 Major quarries in the Blue Lias and their products

Quarry	Locality	Principal quarried formations	Products
ABERTHAW	St Athan ST 040 673	Porthkerry Formation	Cement
RHOOSE	Rhoose ST 066 659	Porthkerry Formation	Cement

important, working a number of coals from the **Pentre** to the **Yard** which, in the south pit, were greatly complicated by the Llanharan Thrust Belt, leading to the eventual closure of the mine in 1962. Cardiff Navigation or Llanely Colliery [ST 0312 8208], which was abandoned in 1929, worked the coals on the downthrow side of the Llanharan Thrust in the Pontyclun area. Earlier attempts to mine coal took place at Llanharry Colliery [ST 0057 8174], which was later absorbed into the opencast site at Trecastle, and at Torgelli [ST 0047 8145], for which no records are available but from which coal extraction is known to have taken place. Trial pits at Llwynmilwas [ST 079 023] were apparently unsuccessful and it can be assumed that recovery was low; in addition, there are numerous bell-pits along the outcrops of the coals.

Coal has been extracted from two opencast sites. The Trecastle site [ST 008 819], worked between 1971 and 1972, extracted 63 000 tonnes from the **Two-Feet Nine**, **Four Feet** and **Abergorki** seams. The Llanilid site [SS 993 819] was opened in 1970 and continues to work the coals from the **Gellideg** to the **Hafod** seams; 3 000 000 tonnes of coking coal had been extracted by 1983.

The coals of the district range from coking coals to bituminous coals, varying in rank from 301 to 801 (NCB classification), with volatile contents of 31 to 35 per cent and low ash values of 3.4 to 11.1 per cent.

IRONSTONE AND IRON ORE

There is evidence that iron ore has been worked in South Wales since Roman times. The early iron industry was based on the clay-ironstones in the Coal Measures, which were generally exploited from surface workings (patchworks). These may be present in the Llanharan and Pontyclun areas, although there are no records of such workings in the district.

The requirements of the South Wales steel industry for high-grade ore resulted in the exploitation of the hematite deposits in the north-east of the district. These extend in a belt between Llanharry [SS 997 807] and Brofiscin [ST 070 819], forming the western end of the Taff's Well – Llanharry ore-field. The earliest workings were from opencast sites at Mwyndy [ST 055 817], Bute [ST 051 817] and Patch Cottages (Llanharry), which are known to have been active prior to 1859. The mines at Trecastle and Mwyndy were in operation as late as 1878 to 1891, but flooding caused their closure shortly afterwards. Most of the deep mining (locally, to depths of 300 m below OD) took place during the 20th century until operations finally ceased at Llanharry in 1976. The deep workings were generally stoped or worked by 'pillar and stall' methods, and were reached by a number of shafts and adits; at Mwyndy and Bute, adits to the deeper ore-bodies extend from the floors of the opencast pits. There are, in addition, a number of trial shafts that failed to prove substantial ore-bodies; these are mostly located in the north-east of the district, but a trial in the north-west near Ty Coch Farm [SS 828 794], may have previously worked small amounts of iron and manganese ore (Watson, 1859).

The combined, post-1859 output from the Bute, Mwyndy and Trecastle mines was 1 661 000 tons of ore. At Llanharry, production rose from an annual figure of 1200 tons in 1901 to 95 000 tons in 1925. Between 1939 and 1945 200 000 tons of ore were produced; in 1967 the figure was 152 000 tonnes, falling to 83 000 tonnes in 1973.

LEAD

Lead ore (galena) was sporadically worked in the district until about 1880 (Lewis, 1967). The most ambitious ventures were at Llangan [SS 9552 7720], where a shaft was sunk to 45 m, Penllyn Court [SS 9677 7688], where 141 tons of ore were produced between 1876 and 1878, and Pontyparc [ST 049 822], where 402 tons were produced between 1757 and 1760. Other mines and surface workings for lead during the 18th and 19th centuries include those at Ogmore Down [SS 885 762], the Golden Mile [SS 942 767], Newton [SS 841 780], Coychurch [SS 940 797] and Merthyr Mawr [SS 883 775]. Shafts were sunk at several other localities, but it is not known if these were trials or if lead was worked from them.

MINOR RESOURCES

Sandstone has been quarried to a limited extent at several localities. The Penarth Group sandstones were previously worked as a source of silica sand and for building stone, at Quarella in Bridgend [SS 904 808], and near Pencoed [SS 948 817]. The Namurian and lower Westphalian sandstones have been quarried for building stone at a few localities along their crop, but may have limited use as a source of aggregate (Wilson and Smith, 1984).

Alluvial gravels were previously worked in limited quantities at Merthyr Mawr [SS 844 771], and deposits of gravelly till were excavated at Brynsadler [ST 029 812], Pontyclun [ST 031 817] and Castell-y-mwnws [ST 021 806].

Glacial silts and clays have been worked in the past for brick manufacture at Pencoed [SS 958 820] and for pottery clay at Ewenny [SS 905 778].

HYDROGEOLOGY

Devonian rocks are of limited potential as aquifers. The mudstone-dominated formations are largely impermeable, and intergranular groundwater movement is restricted in the variably cemented sandstones, the flow occurring mainly through fissures developed in the latter at shallow depths. Most of the outcrop is covered by drift, restricting recharge.

Dinantian limestones form the major aquifer of the district. The limestones have minimal primary permeabilities; however, joints and fissures, particularily in zones of faulting, become enlarged through solution, and groundwater flow is concentrated in these highly permeable zones. The fissures are generally large and groundwater flow can be rapid, issuing from a limited number of large springs the flow-rate of which varies with rainfall. A 6 m-deep collector well with associated springs at Schwyll [SS 888 770] is licensed to abstract the daily equivalent of 355 l/sec. Drilling into the limestones is speculative because the fissure systems tend to be sparsely distributed and not extensively interconnected. However, several deep boreholes have been drilled into the limestones through overlying Mesozoic rocks, with yields from 533 mm diameter boreholes varying from 2.4 l/sec from a 232 m hole in Bridgend [SS 9220 7870], to 17.9 l/sec from a 122 m hole near Coychurch [SS 9350 7934]. Spring water from the limestones is of good quality under low-flow conditions, but after heavy rain may become turbid and polluted. At depth, the water becomes harder; the borehole near Coychurch had a total hardness of 310 mg/l (as calcium carbonate), whilst a 234 m hole at East Aberthaw [ST 0309 6918] had a total hardness of 1172 mg/l.

Silesian strata are unimportant as aquifers in the Bridgend district, for they have a limited outcrop and the dominant lithology is mudstone; groundwater is generally contained in fissures and faults within the sandstones. Mining operations may also influence groundwater movement.

The water-bearing potential of Triassic rocks is restricted to the marginal facies, but is limited by the thickness (generally less than 30 m) and extent of the latter. However, several boreholes drilled into Dinantian limestones at depth also abstract some water from the overlying marginal facies. A well at Ffynnon Fawr [SS 8223 7810] yielded 6.3 l/sec. Water quality is generally good, but hard (up to 300 mg/l total hardness), from shallow sources; at depth however, it tends to become excessively hard and unpotable.

The Lower Lias is a multilayered aquifer, with the limestone beds forming individual aquifers, separated by mudstones through which little groundwater movement occurs. Yields from 100 mm diameter boreholes are generally less than 0.5 l/sec, but a 305 mm diameter borehole, 122 m deep into the Blue Lias at Rhoose Cement Works [SS 0640 6615], yielded 14.1 l/sec. The water is normally of good quality, with total hardness in the range of 250 mg/l to 350 mg/l and a chloride ion concentration of less than 30 mg/l.

The sand and gravel deposits underlying alluvium in the main valleys are saturated and represent a potential source of water, but are not normally utilised as aquifers in this district. A 2.1 m-deep well into till deposits near Hensol Hospital [ST 0509 7895] yielded 2.3 l/sec.

APPENDIX 1

Guide to geological localities along the Heritage Coast near Ogmore-by-Sea

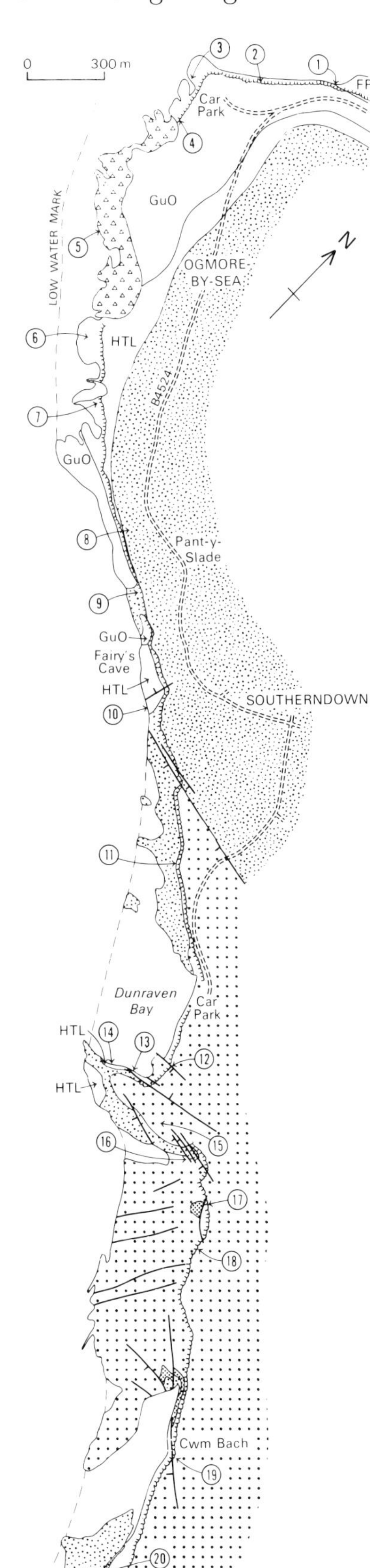

1 Contact between dolomitised Friars Point Limestone and basal unit, Ogwr Member, of the Gully Oolite.

2 Skeletal packstones and ooid/pisoid grainstones of Gully Oolite. Burrows infilled with ooids from overlying bed.

3 Triassic sedimentary 'dykes' of red mudstone and calcareous sandstone infilling fractures in Dinantian limestone.

4 Triassic conglomerates unconformably overlying eroded surface of Dinantian limestone.

5 Triassic conglomerates and breccias cemented by white calcite, pink barites, and galena.

6 Well-bedded skeletal packstones and thin shales of High Tor Limestone. Corals and brachiopods seen on surfaces.

7 Triassic scree deposits unconformably overlying stepped surface of High Tor Limestone.

8 Marginal Lias shelly limestones unconformable on gently dipping High Tor Limestone. Surface of unconformably displays shallow scour channels and borings.

9 Marginal Lias abutting stepped unconformity. Basal breccia cemented, and fossils replaced, by hydrothermal calcite, barytes and galena.

10 Folded Dinantian limestone unconformably overlain by subhorizontal marginal Lias, containing chert fragments.

11 Cliff section displays lateral passage northward from limestone and shales of the Blue Lias into coarse shelly limestones of the marginal facies.

12 4 m-thick Seamouth Limestone (Unit C) overlain by shale-dominated sequence (Unit D).

13 The Dunraven Fault which shows evidence of sinistral strike-slip displacement.

14 Folded and faulted Dinantian High Tor Limestone unconformably overlain by marginal facies of Lower Lias.

15 Vantage point of frontispiece.

16 Folds associated with normal faulting in Liassic limestones and shales.

17 Dome structure at base of cliffs cored by uppermost part of Lavernock Shales.

18 Unit A and most of Unit B of the Porthkerry Formation with the 'Sandwich Bed' near the cliff top.

19 Hanging valley resulting from dissection of glacial valley by cliffline retreat.

20 Marginal Lias, containing fossilised wood fragments, overlain by 4 m of Lavernock Shales, and the Porthkerry Formation. The 'Main Limestone' near the cliff-top.

WARNING: WHEN VISITING THIS AREA BEWARE OF FALLING ROCKS NEAR CLIFFS AND TAKE CARE NOT TO BE CUT OFF BY A RISING TIDE.

APPENDIX 2

Summary of logs of key boreholes

	Depth m
1 Kenfig Borehole [SS 8055 8167]	
Drift	
Glacial sand and gravel	7.32
Mercia Mudstone Group	12.80
'Millstone Grit Series'	48.46
Oystermouth Beds (faulted at base)	69.80
Oxwich Head Limestone	to 91.74
2 BGS Porthcawl Borehole [SS 8112 7773]	
Made ground and weathered solid	3.00
Oxwich Head Limestone	68.88
Pant Mawr Sandstone Member	71.26
Stormy Limestone	132.03
Cornelly Oolite	to 198.31
3 BGS Beacons Down Borehole [SS 8874 7521]	
Made ground	4.44
Cornelly Oolite	7.07
High Tor Limestone	140.33
Gully Oolite	189.14
Friar's Point Limestone	to 196.56
4 BGS Ewenny Borehole [SS 9010 7697]	
Made ground	5.24
Friar's Point Limestone	85.57
Brofiscin Oolite	96.04
Barry Harbour Limestone	144.30
Cwmyniscoy Mudstone	208.67
Castell Coch Limestone	227.70
Tongwynlais Formation	283.31
Quartz Conglomerate Group	to 297.16
5 BGS Glanwenny Borehole [SS 9030 7814]	
Made ground	1.90
Drift	
Till	5.60
Glacial silts and clays	9.95
Porthkerry Formation	10.00
6 Borehole on Bridgend Industrial Estate [SS 9225 7867]	
Drift	
Till	2.7
Lower Lias	152.4
Penarth Group	
Lilstock Formation; marginal facies	158.8
Westbury Formation; sandstone	168.7
Westbury Formation; black shales	169.5
Mercia Mudstone Group; marginal facies	198.9
Lower Limestone Shale Group	to 232.0

	Depth m
7 Borehole on Bridgend Industrial Estate [SS 9350 7934]	
Lower Lias	42.8
Penarth Group	
Lilstock Formation; marginal facies	48.3
Westbury Formation; sandstone	59.6
Westbury Formation; black shales	60.6
Mercia Mudstone Group; marginal facies	91.5
Cornelly Oolite	to 121.9
8 M4 Motorway Borehole [SS 9568 8075]	
Drift	
Alluvium	1.8
River gravels	8.4
Glacial silts and clays	9.1
Till	16.8
Mercia Mudstone Group; marginal facies	17.8
Oxwich Head Limestone	18.6
Pant Mawr Sandstone Member	19.2
Stormy Limestone	to 28.3
9 BGS Pencoed Brick Pit Borehole [SS 9580 8216]	
Made ground	1.3
Drift	
Glacial silts and clays	17.0
10 Penllyn Moor Borehole [SS 9882 7653]	
Drift	13.5
Castell Coch Limestone	27.9
Tongwynlais Formation	to 45.0
11 Borehole at Kingsland [ST 0194 7168]	
St Mary's Well Bay Formation; (?) marginal facies	2.06
Penarth Group	12.35
Mercia Mudstone Group; marginal facies	17.42
12 Aberthaw Borehole [ST 0309 6918]	
Lower Lias and Penarth Group	86.87
Mercia Mudstone Group; marginal facies	167.64
Hunts Bay Oolite Group	to 234.09
13 BGS Pontyclun (Ynys Ddu) Borehole [ST 0344 8191]	
Drift	
Alluvium	1.1
Fluvioglacial sands and gravels	7.75
Glacial silts and clays	11.35
Till	17.60
Middle Coal Measures	to 17.65

APPENDIX 3

BGS publications relevant to the area

Maps at 1:50 000 scale

Bridgend (Sheet 261 and 262) solid and drift. 1990.
Cardiff (Sheet 263) solid. 1968.
Cardiff (Sheet 263) drift. 1990.

Off-shore geology maps at 1:250 000 scale (UTM Series)

51N 04W Bristol Channel, solid geology. 1988.
51N 04W Bristol Channel, sea bed sediments. 1986.

Geophysical maps at 1:250 000 scale (UTM Series)

51N 04W Bristol Channel, aeromagnetic anomaly map. 1980.
51N 04W Bristol Channel, Bouguer gravity anomaly map. 1986.

Hydrogeological maps at 1:125 000 scale

Hydrogeological map of South Wales. 1986

Memoirs

Squirrell, H C, and Downing, R A. 1969. The geology of the South Wales Coalfield, Part I, the country around Newport (Mon.).

Waters, R A, and Lawrence, D J D. 1987. The geology of the South Wales Coalfield, Part III, the country around Cardiff. 3rd edition.

Woodland, A W, and Evans, W B. 1964. The geology of the South Wales Coalfield, Part IV, the country around Pontypridd and Maesteg.

Reports

Harrison, D J, Murray, D W, and Wild, J B L. 1983. Reconnaissance survey of Carboniferous Limestone resources in Wales.

Harrison, D J, 1984. The limestone resources of the Bridgend district—a reconnaissance study.

Wilson, D, and Smith, M. 1985. Planning for development: thematic geology maps, Bridgend area.

Cox, F C, Davies, J R, and Scrivener, R C. 1987. Sources of high grade sandstone for aggregate usage in parts of South West England and South Wales.

1:10 000 Maps

A complete set of 1:10 000 scale geological maps are available for the district with accompanying open-file reports.

The 1:10 000 National Grid maps included in 1:50 000 map Sheets 261 and 262 are listed below with names of the surveyors and dates of survey. Copies of all the maps and open-file reports are available for purchase from the British Geological Survey, Keyworth, Nottingham NG12 5GG.

Map	Surveyor	Date
SS 78 SE	J R Davies	1981
SS 87 NW/ SS 77 NE (part)	J R Davies	1981
SS 87 NE	C J N Fletcher	1981
SS 87 SE/ SS 87 SW (inset)	C J N Fletcher	1982
SS 87 SW (part)	J R Davies, D Wilson	1981, 1983
SS 88 SE (part)	C J N Fletcher	1981, 1983
SS 96 NW	J R Davies	1982
SS 88 SE	C J N Fletcher	1981
SS 96 NE	M Smith	1982
SS 97 NW	D Wilson	1981
SS 97 NE	J R Davies	1983
SS 97 SW	J R Davies, M Smith	1982
SS 97 SE	M Smith	1982
SS 98 SW (part)	J R Davies, M Smith, D Wilson	1981, 1983
SS 98 SE (part)	M Smith	1983
ST 06 NW	C J N Fletcher	1982
ST 06 NE	R A Waters, D Wilson	1976–77, 1982
ST 07 NW	C J N Fletcher	1983
ST 07 NE	K Taylor, D Wilson	1976–77, 1983
ST 07 SW	C J N Fletcher, D Wilson	1982
ST 07 SE	R A Waters, D Wilson	1976–77, 1982
ST 08 SW (part)	M Smith	1983
ST 08 SE (part)	R A Waters, D Wilson	1976, 1983

REFERENCES

Most of the references listed below are held in the Library of the British Geological Survey at Keyworth, Nottingham. Copies of the references can be purchased subject to the current copyright legislation.

Allen, J R L. 1974. The Devonian rocks of Wales and the Welsh Borderland. 47–84 in *The Upper Palaeozoic and post-Palaeozoic rocks of Wales*. Owen, T R (editor). (Cardiff: University of Wales Press.)

— 1975. Source rocks of the Lower Old Red Sandstone: Llanishen Conglomerate of the Cardiff area, South Wales. *Proc. Geol. Assoc.*, Vol. 86, 63–76.

— 1979. Old Red Sandstone facies in external basins, with particular reference to southern Britain. 65–80 *in* The Devonian System, International Symposium on the Devonian System, Bristol, 1978. Scrutton, C T, and Bassett, M G (editors). *Spec. Pap. in Palaeontol.*, No. 23.

Anderson, J G C. 1974. The buried channels, rock-floors and rock-basins and overlying deposits of the South Wales valleys from Wye to Neath. *Proc. S. Wales Inst. Eng.*, Vol. 88, 1–25.

— and Owen, T R. 1968. *The structure of the British Isles*. (Oxford: Pergamon Press.)

Anderton, R, Bridges, P H, Leeder, M R, and Sellwood, B W. 1979. *A dynamic stratigraphy of the British Isles*. (Allen and Unwin.)

Bluck, B J. 1965. The sedimentary history of some Triassic conglomerates in the Vale of Glamorgan, South Wales. *Sedimentology*, Vol. 4, 225–245.

Bott, M H P. 1984. Subsidence mechanisms of Carboniferous Basins of UK. *Eur. Dinant. Envir. 1st Mtg Abstr.* No. 96, (Department of Earth Sciences, Open University.)

Bowen, D Q. 1970. South-east and central South Wales. 197–227 in *The glaciations of Wales and adjoining regions*. Lewis, C A (editors). (London: Longman.)

— 1973a. The Pleistocene history of Wales and the borderland. *Geol. J.*, Vol. 8, 207–224.

— 1973b. The Pleistocene succession of the Irish Sea. *Proc. Geol. Assoc.*, Vol. 84, 249–269.

— 1974. The Quaternary of Wales. 373–426 in *The Upper Palaeozoic and post-Paleozoic rocks of Wales*. Owen, T R (editor). (Cardiff: University of Wales Press.)

— 1981. The 'South Wales end-moraine': fifty years after. 60–67 in *The Quaternary in Britain*. Neale, J, and Flenley, J (editors). (Oxford: Pergamon Press.)

Burchette, T P. 1981. The Lower Limestone Shales. 13–27 in *A field guide to the Carboniferous Limestone around Abergavenny*. Wright, V P, Raven, M, and Burchette, T P. (Department of Geology, University College of Wales, Cardiff.)

— and Riding, R. 1977. Attached vermiform gastropods in Carboniferous marginal marine stromatolites and biostromes. *Lethaia*, Vol. 10, 17–28.

Charlesworth, J K. 1929. The South Wales end-moraine. *Q. J. Geol. Soc. London*, Vol. 85, 335–358.

Cope, J C W, Getty, T A, Howarth, M K, Morton, N, and Torrens, H S. 1980. A correlation of Jurassic rocks in the British Isles. Part One: Introduction and Lower Jurassic. *Spec. Rep. Geol. Soc. London*, No. 14.

Crampton, C B. 1960. Analysis of heavy minerals in the Carboniferous Limestone, Millstone Grit and soils derived from certain glacial gravels of Glamorgan and Monmouth. *Trans. Cardiff Naturalists Soc.*, Vol. 87, 13–22.

— 1961. An interpretation of the micromineralogy of certain Glamorgan soils: the influence of ice and wind. *J. Soil Sci.*, Vol. 12, 158–171.

— 1964. Certain aspects of soils developed on calcareous parent materials in South Wales. *Trans. Cardiff Naturalists Soc.*, Vol. 91, 4–16.

— 1966. Certain effects of glacial events in the Vale of Glamorgan, South Wales. *J. Glaciology*, Vol. 6, 261–266.

Dixey, F, and Sibly, T F. 1918. The Carboniferous Limestone Series on the south-eastern margin of the South Wales Coalfield. *Q. J. Geol. Soc. London*, Vol. 73, 111–164.

Dixon, E E L, and Vaughan, A. 1912. The Carboniferous succession in Gower (Glamorganshire). *Q. J. Geol. Soc. London*, Vol. 67, 477–571.

Dott, R H Jr, and Bourgeoise, J. 1982. Hummocky stratification: Significance of its variable bedding sequences. *Bull. Geol. Soc. Am.*, Vol. 93, 663–680.

Dunning, F W. 1977. Caledonian–Variscan relations in northwest Europe. 165–180 in *La Chaine Varisque d'Europe Moyenne et Occidentale*, Editions du CNRS, No. 243. (Paris: CNRS)

Elliott, T, and Ladipo, K. 1981. Syn-sedimentary gravity slides (growth faults) in the Coal Measures of South Wales. *Nature, London*, Vol. 291, 220–222.

Fletcher, C J N. 1988. Tidal erosion, solution cavities and exhalative mineralisation associated with the Jurassic unconformity at Ogmore, South Glamorgan. *Proc. Geol. Assoc.*, Vol. 99, 1–14.

Fletcher, C J N, and Young, B R. 1987. Rosasite from Bute Quarry, Mid-Glamorgan. The first reported occurrence in Wales. *J. Russell Soc.*, Vol. 2, 1–7.

Francis, E H. 1959. The Rhaetic of the Bridgend district. *Proc. Geol. Assoc.*, Vol. 70, 158–170.

Freshney, E C, and Taylor, R T. 1980. The Variscides of southwest Britain. 49–57 in *United Kingdom: introduction to general geology and guides to excursions 002, 055, 093 and 151*. Owen, T R (editor). 26th International Geological Congress, Paris, 1980. (London. Institute of Geological Sciences.)

Gayer, R A, Allen, K C, Bassett, M G, and Edwards, D. 1973. The structure of the Taff Gorge area, Glamorgan, and the stratigraphy of the Old Red Sandstone-Carboniferous Limestone transition. *Geol. J.*, Vol. 8, 345–374.

— and Criddle, A. J. 1970. Mineralogy and genesis of the Llanharry iron ore deposits, Glamorgan. 605–626 in Mining and petroleum. *Proc. 9th Commonwealth Min. and Metallurgical Congress 1969*, Vol. 2. (London: Institution of Mining and Metallurgy.)

George, T N. 1933. The Carboniferous Limestone Series in the west of the Vale of Glamorgan. *Q. J. Geol. Soc. London*, Vol. 89, 221-271.

— 1955. The Namurian Usk Anticline. *Proc. Geol. Assoc.*, Vol. 66, 297-316.

— 1958. Lower Carboniferous palaeogeography of the British Isles. *Proc. Yorkshire Geol. Soc.*, Vol. 31, 227-318.

— 1970. British regional geology; South Wales (3rd edition). (London: HMSO for Institute of Geological Sciences.)

— 1974. Lower Carboniferous rocks in Wales. 85-115 in *The Upper Palaeozoic and post-Palaeozoic rocks in Wales.* Owen, T R (editor). (Cardiff: University of Wales Press.)

— 1978. Eustasy and tectonics: sedimentary rhythms and stratigraphical units in British Dinantian correlation. *Proc. Yorkshire Geol. Soc.*, Vol. 42, 229-262.

— Johnson, G A L, Mitchell, M, Prentice, J E, Ramsbottom, W H C, Sevastopulo, G D, and Wilson, R B. 1976. A correlation of Dinantian rocks in the British Isles. *Spec. Rep. Geol. Soc. London*, No. 7.

Ginsberg, R N. 1975. *Tidal deposits.* (New York: Springer-Verlag.)

— and James, N P. 1974. Holocene carbonate sediments of continental shelves. 137-157 in *Continental margins.* Burk, C A, and Drake, C L (editors). (New York: Springer-Verlag.)

Goldring, R, and Bridges, P H. 1973. Sublittoral sheet sandstones. *J. Sediment. Petrol.*, Vol. 43, 736-747.

Griffiths, J C. 1939. The mineralogy of the glacial deposits of the region between the rivers Neath and Towy, South Wales. *Proc. Geol. Assoc.*, Vol. 50, 433-462.

Hallam, A. 1960. A sedimentary and faunal study of the Blue Lias of Dorset and Glamorgan. *Philos. Trans. R. Soc. London, Ser. B*, Vol. 243, 1-14.

— 1964. Origin of the limestone-shale rhythm in the Blue Lias of England: a composite theory. *J. Geol.*, Vol. 72, 157-169.

Hancock, P L, Dunne, W M, and Tringham, M E. 1983. Variscan deformation in southwest Wales. 47-73 in *The Variscan fold belt in the British Isles.* Hancock, P L (editor). (Bristol: Adam Hilger.)

Harrison, D J. 1984. *The limestone resources of the Bridgend area—a reconnaissance survey. Description of 1:50 000 geological sheet 262.* (Keyworth: Institute of Geological Sciences.)

Hird, K, Tucker, M E, and Waters, R A. 1987. Petrography, geochemistry and origin of Dinantian dolomites from south-east Wales. 359-378 in *European Dinantian environments.* Miller, J, Adams, A E, and Wright, V P (editors). *Geol. J. Spec. Issue* No. 12.

Hodges, P. 1986. The Lower Lias (Lower Jurassic) of the Bridgend area, South Wales. *Proc. Geol. Assoc.*, Vol. 93, 237-242.

Howard, F T, and Small, E W. 1901. Notes on ice action in South Wales. *Trans. Cardiff Naturalists Soc.*, Vol. 32, 44-48.

Ivimey-Cook, H C. 1974. The Permian and Triassic deposits of Wales. 295-321, in *The Upper Palaeozoic and post-Palaeozoic rocks of Wales.* Owen, T R (editor). (Cardiff: University of Wales Press.)

Kellaway, G A, and Welch, F B A. 1955. The Upper Old Red Sandstone and Lower Carboniferous rocks of Bristol and the Mendips compared with those of Chepstow and the Forest of Dean. *Bull. Geol. Surv. G.B.*, No. 9, 1-21.

Kelling, G. 1974. Upper Carboniferous sedimentation in Wales. 185-224 in *The Upper Palaeozoic and post-Palaeozoic rocks of Wales.* Owen, T R (editor). (Cardiff: University of Wales Press.)

Klappa, C F. 1980. Rhizoliths in terrestrial carbonates: classification, recognition, genesis and significance. *Sedimentology*, Vol. 27, 613-629.

Lawrence, D J D, and Waters, R A. 1978. St Fagans Borehole. 10-11 *in* IGS Boreholes, 1977. *Rep. Inst. Geol. Sci.*, No. 78/21.

Leeder, M R. 1976. Sedimentary facies and the origins of basin subsidence along the northern margin of the supposed Hercynian ocean. *Tectonophysics*, Vol. 36, 167-179.

— 1982. Upper Palaeozoic basins of the British Isles—Caledonide inheritance versus Hercynian plate margin processes. *J. Geol. Soc. London*, Vol. 139, 479-491.

Lewis, W J. 1967. *Lead mining in Wales.* (Cardiff: University of Wales Press.)

Matthews, S C. 1974. Exmoor thrust? Variscan front? *Proc. Ussher. Soc.*, Vol. 3, 82-94.

Mayall, M J. 1983. An earthquake origin for synsedimentary deformation in a late Triassic (Rhaetian) lagoonal sequence, south-west Britain. *Geol. Mag.*, Vol. 120, 613-622.

Mitchell, G F. 1960. The Pleistocene history of the Irish Sea. *Adv. Sci.*, Vol. 17, 313-325.

Orbell, G. 1973. Palynology of the British Rhaeto-Liassic. *Bull. Geol. Surv. G.B*, Vol. 44, 1-44.

Owen, T. R. 1964. The tectonic framework of Carboniferous sedimentation in South Wales. 301-307 in *Deltaic and shallow marine deposits.* Van Straaten, L M J V (editor). (Amsterdam: Elsevier.)

— 1967. 'From the South': a discussion of two recent papers in the proceedings. *Proc. Geol. Assoc.*, Vol. 78, 595-599.

— 1971. The relationship of Carboniferous sedimentation to structure in South Wales. *C. R. 6me. Congr. Strat. Carb. III*, 1305-1316.

— 1974. The Variscan orogeny in Wales. 285-294 in *The Upper Palaeozoic and post-Palaeozoic rocks of Wales.* Owen, T R (editor). (Cardiff: University of Wales Press.)

— and Weaver, J D. 1983. The structure of the main South Wales Coalfield and its margins. 75-87 in *The Variscan fold belt in the British Isles.* Hancock, P L (editor). (Bristol: Adam Hilger.)

Perkins, J W, Gayer, R A, Baker, J W, and Williams, G D. 1979. *The Glamorgan Heritage Coast: A guide to its geology. (Bridgend: The Glamorgan Heritage Coast Joint Management and Advisory Committee.)*

Ramsbottom, W H C. 1954. In *Summary of progress of the Geological Survey of Great Britain for 1953.* (London: HMSO.)

— 1973. Transgressions and regressions in the Dinantian: a new synthesis of British Dinantian stratigraphy. *Proc. Yorkshire Geol. Soc.*, Vol. 39, 567-607.

— 1977. Major cycles of transgression and regression (mesothems) in the Namurian. *Proc. Yorkshire Geol. Soc.*, Vol. 41, 261-291.

— 1979. Rates of transgression and regression in the Carboniferous of NW Europe. *J. Geol. Soc. London*, Vol. 136, 147-153.

— Calver, M A, Eagar, R M C, Hodson, F, Holliday, D W, Stubblefield, C J, and Wilson, R B. 1978. A correlation of Silesian rocks in the British Isles. *Spec. Rep. Geol. Soc. London*, No. 10.

Richardson, L. 1905. The Rhaetic and contiguous deposits of Glamorganshire. *Q. J. Geol. Soc. London*, Vol. 61, 385-424.

— 1911. The Rhaetic and contiguous deposits of West, Mid and part of East Somerset. *Q. J. Geol. Soc. London*, Vol. 67, 1-72.

RIDING, R, and WRIGHT, V P. 1981. Paleosols and tidal-flat/lagoon sequences on a Carboniferous carbonate shelf: sedimentary associations of triple disconformities. *J. Sediment. Petrol.*, Vol. 51, 1323–1339.

SCHOLLE, P A, BEBOUT, D G, and MOORE, C H (editors). 1983. Carbonate depositional environments. *Mem. Am. Assoc. Pet. Geol.*, No. 33.

SELLWOOD, B W. 1970. The relation of trace fossils to small scale sedimentary cycles in the British Lias. 489–504 in *Trace fossils*. CRIMES, T P, and HARPER, J C (editors). (Liverpool: Seel House Press.)

— DURKIN, M K, and KENNEDY, W J. 1970. Field meeting on the Jurassic and Cretaceous rocks of Wessex. *Proc. Geol. Assoc.*, Vol. 81, 715–732.

— and JENKYNS, H C. 1975. Basins and swells and the evolution of an epeiric sea (Pliensbachian–Bajocian of Great Britain). *J. Geol. Soc. London*, Vol. 131, 373–388.

SHACKLETON, R M. 1984. Thin skinned tectonics, basement control and the Variscan front. 125–129 in *Variscan tectonics of the North Atlantic Region*. HUTTON, D H W, and SANDERSON, D J (editors). *Spec. Publ. Geol. Soc. London*, No. 14.

— RIES, A C, and COWARD, M P. 1982. An interpretation of the Variscan structures in SW England. *J. Geol. Soc. London*, Vol. 139, 533–541.

SOMERVILLE, I D, and STRANK, A R E. 1984. Discovery of Arundian and Holkerian faunas from a Dinantian platform succession in North Wales. *Geol. J.*, Vol. 19, 85–104.

SQUIRRELL, H C, and DOWNING, R A. 1969. Geology of the South Wales Coalfield, Part I, the country around Newport (Mon). *Mem. Geol. Surv. G.B.*, Sheet 249.

— and TUCKER, E V. 1960. The geology of the Woolhope Inlier (Herefordshire). *Q. J. Geol. Soc.* London, Vol. 116, 139–181.

STRAHAN, A, and CANTRILL, T C. 1904. The geology of the South Wales Coalfield, Part VI, the country around Bridgend. *Mem. Geol. Surv. G.B*, Sheet 261/262.

STRANK, A R E. 1982. Asbian and Brigantian foraminifera from the Beckermonds Scar Borehole. *Proc. Yorkshire Geol. Soc.*, Vol. 44, 103–108.

TAWNEY, E B. 1866. On the Western Limit of the Rhaetic Beds in South Wales, and on the position of the Sutton Stone. *Q. J. Geol. Soc. London*, Vol. 22, 69–93.

THOMAS, T M. 1953. New evidence of intraformational piping at two separate horizons in the Carboniferous Limestone (D_2) at South Cornelly, Glamorgan. *Geol. Mag.*, Vol. 93, 73–82.

TRENHAILE, A S. 1971. Lithological control of high water ledges in the Vale of Glamorgan, Wales. *Geog. Annlr.*, Vol. 53A, 56–59.

TRUEMAN, A E. 1920. The Liassic rocks of the Cardiff district. *Proc. Geol. Assoc.*, Vol. 31, 93–107.

— 1922. The Liassic rocks of Glamorgan. *Proc. Geol. Assoc.*, Vol. 33, 245–284.

— 1930. The Lower Lias (*bucklandi* Zone) of Nash Point, Glamorgan. *Proc. Geol. Assoc.*, Vol. 41, 148–159.

TUCKER, M E. 1977. The marginal Triassic deposits of South Wales: continental facies and palaeogeography. *Geol. J.*, Vol. 12, 169–188.

— 1978. Triassic lacustrine sediments from South Wales: shore-zone clastics, evaporites and carbonates. 205–224 in *Modern and ancient lake sediments*. MATTER, A, and TUCKER, M E (editors). *Spec. Publ. Int. Assoc. Sedimentologists*, No. 2.

TUNBRIDGE, I P. 1986. Mid-Devonian tectonics and sedimentation in the Bristol Channel area. *J. Geol. Soc. London*, Vol. 143, 107–115.

VAUGHAN, A. 1905. The palaeontological sequence in the Carboniferous Limestone of the Bristol area. *Q. J. Geol. Soc. London*, Vol. 61, 181–305.

WALKDEN, G M. 1974. Palaeokarstic surfaces in Upper Viséan (Carboniferous) Limestones of the Derbyshire Block, England. *J. Sediment. Petrol.*, Vol. 44, 1232–1247.

WALMSLEY, V G. 1959. The geology of the Usk Inlier, Monmouthshire. *Q. J. Geol. Soc. London*, Vol. 114, 483–521.

WARRINGTON, G, AUDLEY-CHARLES, M G, ELLIOTT, R E, EVANS, W B, IVIMEY-COOK, H C, KENT, P, ROBINSON, P L, SHOTTON, F W, and TAYLOR, F M. 1980. A correlation of Triassic rocks in the British Isles. *Spec. Rep. Geol. Soc. London*, No. 13.

WATERS, R A. 1984. Some aspects of the Black Rock Limestone and Gully Oolite (Dinantian) in the eastern Vale of Glamorgan. *Proc. Geol. Assoc.*, Vol. 95, 391–392.

— and LAWRENCE, D J D. 1987. Geology of the South Wales Coalfield, Part III, the country around Cardiff (3rd edition). *Mem. Br. Geol. Surv.*, Sheet 263 (England and Wales).

WATSON, J J W. 1859. The haematitic deposits of Glamorganshire. *The Geologist*, Vol. 2, 241–256.

WHITTAKER, A. 1975. A postulated post-Hercynian rift valley system in southern Britain. *Geol. Mag.*, Vol. 112, 137–149.

— 1978. The lithostratigraphical correlation of the uppermost Rhaetic and lowermost Liassic strata of the west Somerset and Glamorgan areas. *Geol. Mag.*, Vol. 115, 63–67.

WILLIAMS, A, CALDWELL, N, and DAVIES, P. 1981. *The Glamorgan Heritage Coast: A guide to its landforms. Bridgend: The Glamorgan Heritage Coast Joint Management and Advisory Committee, Bridgend.)*

WILLIAMS, G D, and CHAPMAN, T J. 1986. The Bristol–Mendip foreland thrust belt. *J. Geol. Soc. London*, Vol. 143, 63–73.

WILLIAMS, M. 1958. Geology and mineralisation at the Llanharry hematite deposit, South Wales. Unpublished PhD thesis, Department of Mining Geology, Imperial College, London.

WILSON, D, and SMITH, M. 1985. *Planning for development: thematic geology maps, Bridgend area. Geological report for DOE.* (Aberystwyth: British Geological Survey.)

WOBBER, F J. 1965. Sedimentology of the Lias (Lower Jurassic) of South Wales. *J. Sediment. Petrol.*, Vol. 35, 683–703.

— 1966. A study of the depositional area of the Glamorgan Lias. *Proc. Geol. Assoc.*, Vol. 77, 127–137.

— 1968. A faunal analysis of the Lias (Lower Jurassic) of South Wales (Great Britain). *Palaeogeogr. Palaeoclimatol. Palaeoecol.*, Vol. 10, 269–308.

WOODLAND, A W, EVANS, W B, and STEPHENS, J V. 1957a. Classification of the Coal Measures of South Wales with special reference to the upper Coal Measures. *Bull. Geol. Surv. G.B*, No. 13, 6–13.

— ARCHER, A A, and EVANS, W B. 957b. Recent boreholes into the Lower Coal Measures below the Gellideg-Lower Pumpquart Coal horizon in South Wales. *Bull. Geol. Surv. G.B*, No. 13, 39–60.

— and EVANS, W B. 1964. The geology of the South Wales Coalfield, part IV, the country around Pontypridd and Maesteg. *Mem. Geol. Surv. G.B*, Sheet 248.

GLOSSARY

Autobrecciation A fragmentation process operating penecontemporaneously with deposition of the rock unit.

Beekite Botryoidal, commonly concentric, accretions of white, opaque silica commonly forming bands or layers on silicified fossils.

Bioclastic A term used to describe material consisting of fragmental remains of organisms, commonly shells (see also **Skeletal**).

Bioherm A mound, dome, lens or reef-like mass, built up by sedentary organisms and largely composed of their remains (see also **Thrombolitic**).

Bioturbation The churning and stirring of a sediment by organisms.

Bird's-eye structures Small patches of sparry calcite, occurring as irregular or elongate spots, blebs and tubes, commonly aligned parallel to bedding; they generally represent shrinkage cracks, resulting from desiccation.

Calcarenite A limestone consisting predominantly of detrital sand-sized particles of calcium carbonate (lithoclasts, skeletal grains, ooids etc.).

Calcite mudstone A fine-grained limestone with less than 10 per cent of the particles greater than 20 μ diameter.

Calcrete Secondary accumulations of calcium carbonate within soil profiles.

Coprolites The fossilised faeces of vertebrates, generally phosphatic.

Coquinoid A term referring to limestones mainly composed of mechanically sorted, and often unbroken, shell material.

Cryptalgal lamination A distinctive lamination, believed to record successive growths of blue-green algal mats (see also **Stromatolite**).

Fenestrae (-al) see Bird's-eye structures.

Flaser-bedded Ripple cross-laminated sandstone sequences in which mud occurs as discontinuous streaks and lenses preserved in the troughs between ripple crests (see also **Linsen bedded**).

Grainstone A textural term applied to limestones in which the constituent grains form a self-supporting framework, in which the interstices are free of mud-grade carbonate (cf. **Packstone**).

Growth fault A fault in a sedimentary sequence which forms contemporaneously with deposition, such that the throw increases with depth and the strata on the downthrown side are thicker than their correlatives on the upthrown side.

Interstadial A climatic amelioration within a glaciation during which there was stillsand or temporary retreat of the ice.

Intraclasts Redeposited fragments of previously lithified limestone of local derivation.

Intraformational conglomerate A conglomerate composed of clasts derived from subjacent parts of the formation in which it oocurs.

Kettle hole A bowl-shaped depression in glacial deposits, formed by the melting of a detached block of stagnant ice.

Linsen-bedded A mudstone-dominated sequence with lenses of ripple cross-laminated sandstone.

Micrite A term partly synonymous with calcite mudstone (see above) but used here also to describe a fine-grained carbonate cement.

Ooid A spherical or subspherical grain, less than 2 mm diameter, composed most commonly of concentric layers of fine-grained fibrous calcite surrounding a nucleus; an oolite is a rock composed dominantly of ooids.

Ossiferous Rich in bone fragments.

Packstone A textural term applied to limestones in which the constituent grains form a self-supporting framework and in which the interstices are wholly or partly infilled with carbonate mud (cf. **Grainstone**).

Palaeokarstic surface An ancient karstic surface; karstification is a weathering effect in limestone terrains, in which dissolution processes play a dominant role.

Paralic A term applied to sedimentary sequences which are transitional between marine and nonmarine settings.

Paratectonic A term generally applied to the low-grade metamorphic or nonmetamorphic parts of an orogenic belt.

Pedogenic A term used to describe processes involved in soil formation.

Peloid A grain composed of fine-grained calcareous material (micrite), commonly rounded but irrespective of size or origin.

Pisolitic A rock composed of pisoids, which are grains, similar in structure and composition to ooids, but larger (over 2 mm diameter) and less regular in shape.

Pseudobrecciation An effect of recrystallisation, imparting a breccia-like appearance to limestones.

Skeletal A term referring to complete or fragmented shell material (see also **Bioclastic**).

Stromatolite An organo-sedimentary structure in which the sediment is trapped by the successive growth and binding action of blue-green algal mats (see also **Cryptalgal lamination**).

Terra fusca A brown carbonate-rich loam, commonly formed in limestone terrains under humid, warm temperate climates.

Terra rossa A reddish brown residual soil, commonly formed in limestone terrains under warm temperate to subtropical (Mediterranean-type) climates.

Thrombolitic A term applied to a cryptalgal structure, like a stromatolite, but with an obscurely clotted, rather than laminated, internal structure.

Till Unsorted and unstratified material, deposited directly from a glacier without any subsequent reworking by meltwater; boulder clay is a term commonly used as an equivalent of till, but should be restricted to till deposits comprising clasts of various sizes in a stiff clay matrix.

Wackestone A textural term applied to limestones in which the constituent grains over 20 μ are carbonate mud-supported and form over 10 per cent of the rock.

INDEX

Italics figures refer to tables and illustrations

BRITISH GEOLOGICAL SURVEY

Keyworth, Nottingham NG12 5GG
(06077) 6111

Murchison House, West Mains Road,
Edinburgh EH9 3LA   031-667 1000

London Information Office, Natural History Museum
Earth Galleries, Exhibition Road, London SW7 2DE
071-589 4090

The full range of Survey publications is available through the Sales Desks at Keyworth, Murchison House, Edinburgh, and at the BGS London Information Office in the Natural History Museum, Earth Galleries. The adjacent bookshop stocks the more popular books for sale over the counter. Most BGS books and reports are listed in HMSO's Sectional List 45, and can be bought from HMSO and through HMSO agents and retailers. Maps are listed in the BGS Map Catalogue and the Ordnance Survey's Trade Catalogue, and can be bought from Ordnance Survey agents as well as from BGS.

The British Geological Survey carries out the geological survey of Great Britain and Northern Ireland (the latter as an agency service for the government of Northern Ireland), and of the surrounding continental shelf, as well as its basic research projects. It also undertakes programmes of British technical aid in geology in developing countries as arranged by the Overseas Development Administration.

The British Geological Survey is a component body of the Natural Environment Research Council.

Maps and diagrams in this book use topography based on Ordnance Survey mapping

HMSO publications are available from:

HMSO Publications Centre
(Mail and telephone orders)
PO Box 276, London SW8 5DT
Telephone orders 071-873 9090
General enquiries 071-873 0011
Queueing system in operation for both numbers

HMSO Bookshops
49 High Holborn, London WC1V 6HB
071-873 0011 (Counter service only)
258 Broad Street, Birmingham B1 2HE
021-643 3740
Southey House, 33 Wine Street, Bristol BS1 2BQ
(0272) 264306
9 Princess Street, Manchester M60 8AS
061-834 7201
80 Chichester Street, Belfast BT1 4JY
(0232) 238451
71 Lothian Road, Edinburgh EH3 9AZ
031-228 4181

HMSO's Accredited Agents
(see Yellow Pages)

And through good booksellers